Heike Hufnagel

A Probabilistic Framework for Point-Based Shape Modeling
in Medical Image Analysis

VIEWEG+TEUBNER RESEARCH

Medizintechnik – Medizinische Bildgebung, Bildverarbeitung und bildgeführte Interventionen

Herausgegeben von Prof. Dr. Thorsten M. Buzug,
Institut für Medizintechnik, Universität zu Lübeck

Die medizinische Bildgebung erforscht, mit welchen Wechselwirkungen zwischen Energie und Gewebe räumlich aufgelöste Signale von Zellen oder Organen gewonnen werden können, die die Form oder Funktion eines Organs charakterisieren. Die Bildverarbeitung erlaubt es, die in physikalischen Messsignalen und gewonnenen Bildern enthaltene Information zu extrahieren, für den Betrachter aufzubereiten sowie automatisch zu interpretieren. Beide Gebiete stützen sich auf das Zusammenwirken der Fächer Mathematik, Physik, Informatik und Medizin und treiben als Querschnittsdisziplinen die Entwicklung der Gerätetechnologie voran.
Die Reihe Medizintechnik bietet jungen Wissenschaftlerinnen und Wissenschaftlern ein Forum, ausgezeichnete Arbeiten der Fachöffentlichkeit vorzustellen, sie steht auch Tagungsbänden offen.

Heike Hufnagel

A Probabilistic Framework for Point-Based Shape Modeling in Medical Image Analysis

With a foreword by Prof. Dr. rer. nat. Heinz Handels

VIEWEG+TEUBNER RESEARCH

Bibliographic information published by the Deutsche Nationalbibliothek
The Deutsche Nationalbibliothek lists this publication in the Deutsche Nationalbibliografie;
detailed bibliographic data are available in the Internet at http://dnb.d-nb.de.

Dissertation Universität zu Lübeck, 2010

1st Edition 2011

Cover design: KünkelLopka Medienentwicklung, Heidelberg
Printed on acid-free paper

ISBN 978-3-8348-1722-8

To Emmi and Evi

"Die gefährlichste aller Weltanschauungen ist die Weltanschauung der Leute, welche die Welt nie angeschaut haben."
(The most dangerous of all world-views is the one of people who have never viewed the world.)

Zugeschrieben: Alexander von Humboldt

Preface

The book A Probabilistic Framework for Point-Based Shape Modeling in Medical Image Analysis by Dr. Heike Hufnagel is the third volume of the new Vieweg+Teubner series of excellent theses in medical engineering. The thesis of Dr. Hufnagel has been selected by an editorial board of highly recognized scientists working in that field.

The Vieweg+Teubner series aims to establish a well defined forum for monographs and proceedings on medical engineering. The series publishes works that give insights into the novel developments with emphasis on imaging, image computing and image-guided intervention.

Prospective authors may contact the series editor about future publications within the series at:

Prof. Dr. Thorsten M. Buzug
Series Editor Medical Engineering

Institute of Medical Engineering
University of Lübeck
Ratzeburger Allee 160
23562 Lübeck
Web: www.imt.uni-luebeck.de
Email: buzug@imt.uni-luebeck.de

Foreword

Medical image segmentation is a key problem in the field of medical image computing. The objective of the segmentation process is the localization and delineation of relevant structures like tumors, organs, vessels, bones etc. in medical image data. Segmentations are needed to generate patient-specific 3D models of organs and tumors and other structures of interest e.g. for improved radiation therapy planning or in computer and robot assisted surgery. Furthermore, many quantitative parameters can only be extracted after segmenting an object. Beside the object volume, further quantitative image parameters to characterize the state and changes of image structures of interest (e.g. tumors, organs, vessels, bones etc.) can be extracted in a reproducible and objective way.

Although many approaches for the segmentation of medical images have been developed in the past years, automatic image segmentation is still a difficult task and further developments are needed to increase the grade of automation, accuracy, reproducibility, and robustness. The experience shows that the segmentation quality of pure data-driven approaches is often insufficient. This is due to low contrasts between neighboring image objects, imaging-related problems like noise or artifacts, and to the anatomical variability of organs and tissues. Here, model-based image segmentation algorithms as described in this book are of growing interest, because they allow to incorporate a-priori knowledge about image structures into the segmentation process.

In this book, Heike Hufnagel considers extended statistical shape models as a powerful tool to model organ shapes and to improve the 3D segmentation of modeled organs by using shape information. Shape models provide a statistical representation of object shapes distributions generated from a set of segmented training image data. Here, Heike Hufnagel addresses a central problem: Standard approaches to generate statistical shape models postulate a unique point-to-point correspondence between the surface points of different organs, which often cannot be defined clearly due to uncertainties of the shape surface representations. To model these uncertainties in an adequate way, she develops a novel statistical shape model using correspondence probabilities instead of 1-to-1 correspondences. This approach offers the possibility to model the shape and its variations of organs in a new way and with high accuracy. This extension leads to a new framework for the generation of statistical shape models with probabilistic point correspondences and their use during the segmentation process. In the second part of the book, Heike Hufnagel develops an elegant and concise mathematical framework based on level sets which allows to employ the probabilistic shape model to guide and to restrict the 3D segmentation of organs in medical image data like CT or MRT data sets.

The reader of this book will get a deep insight into the field of statistical shape models, into their mathematical background as well as into the algorithms for model based image segmentation of organs. Many medical examples illustrate the methods and give the reader an impression of the application of the methods in practice. This book is of high value for all readers interested in shape modelling of organs and model based segmentation of medical image data.

Prof. Dr. rer. nat. habil. Heinz Handels
Institute of Medical Informatics
University of Lübeck

Acknowledgments

Before starting this thesis I did not know what I would get myself into, and when I finally realized it and it was too late, I was very glad to discover that I did not have to walk this path alone.

To begin with, I would like to thank my director and doctoral advisor Heinz Handels for offering me the opportunity and the work environment for my research at the IMI, and I also thank him greatly for his trust in my capabilities, his good advice and constructive discussions.

I thank my director Nicholas Ayache from INRIA for kindly integrating me into his team, for supporting my work and giving me direction.

My deep gratitude goes to my advisors, they gave me inspiration, did not avoid heated discussions and taught me a lot about computer science and the world of research in general: Xavier Pennec who guided my exploration of the fascinating realms of mathematics and Jan Ehrhardt who managed the precarious balance between supervisor and dear friend.

I also want to take this opportunity to thank Bernd Fischer who has always accompanied my work from afar for his valuable comments. I thank Tobias Heimann for enthusiastic discussions about shape and his kind cooperation.

From all my heart I thank Ender Konukoglu and Andrea Martini who still believed in me in times when I did not anymore and who I always could rely on for scientific and emotional support. My life would be a lot more confused without them.

With Florence Billet and Jean-Marc Peyrat I walked the same path throughout, and I cannot thank them enough for their empathic companionship which made everything so much easier.

In both my teams I was lucky to meet great assistants: I thank Isabelle Strobant and Renate Reche for efficiently simplifying administrative matters and for having an open ear for all kind of problems.

I thank all my colleagues from the Asclepios team: open doors and the Queen's leg always invited fruitful scientific discussions but also - maybe even more importantly - valuable social and cultural exchange in an international environment. I especially thank Marius Linguraru, Tom Vercauteren, Mauricio Reyes-Aguirre, Olivier Clatz, and Jimena Costa for ties beyond academic issues.

Profoundly I thank my colleagues at the IMI who became my friends. Apart from the scientific support and encouragement they offered, they were the reason I always liked going to work even in difficult too-close-to-deadline or hidden-program-bug times: I thank René Werner for letting me bathe in his serenity, Alex Schmidt-Richberg for forcing his programming skills upon me and for his kindness, Nils Forkert for philosophical (cigarette) breaks even in the middle of the night, Dennis Säring for down-to-earth words at the right time, and Matthias Färber for lightening my view on things. For his patience and willingness to help me with technical computer matters I thank Martin Riemer.

I thank Ricardo Martinez for sharing my life and making it more adventurous and whole, and I thank my parents and my brother for still being my home when needed.

I thank also the German Academic Exchange Service (DAAD) and the German Research Foundation (DFG) that supported my research financially.

Heike Hufnagel

Abstract

This thesis centers on the development of a point-based statistical shape model relying on correspondence probabilities in a sound mathematical framework. Further focus lies on the integration of the model into a segmentation method where a novel approach is taken by combining an explicitly represented shape prior with an implicitly represented segmentation contour.

In medical image analysis, the notion of shape is recognized as an important feature to distinguish and analyse anatomical structures. The modeling of shape realized by the concept of statistical shape models constitutes a powerful tool to facilitate the solutions to analysis, segmentation and reconstruction problems. A statistical shape model tries to optimally represent a set of segmented shape observations of any given organ via a mean shape and a variability model. A fundamental challenge in doing statistics on shapes lies in the determination of correspondences between the shape observations. The prevailing assumption of one-to-one point correspondences seems arguable due to uncertainties of the shape surface representations as well as the general difficulty of pinpointing exact correspondences.

In this thesis, the following solution to the point correspondence problem is derived: For all point pairs, a correspondence probability is computed which amounts to representing the shape surfaces by Mixtures of Gaussians. This approach allows to formulate the model computation in a generative framework where the shape observations are interpreted as randomly generated by the model. Based on that, the computation of the model is then treated as an optimization problem. An algorithm is proposed to optimize for model parameters and observation parameters through a single maximum a posteriori criterion which leads to a mathematically sound and unified framework.

The method is evaluated and validated in a series of experiments on synthetic and real data. To do so, adequate performance measures and metrics are defined based on which the quality of the new model is compared to the qualities of a classical point-based model and of an established surface-based model that both rely on one-to-one correspondences.

A segmentation algorithm is developed which employs the a priori shape knowledge inherent in the statistical shape model to constrain the segmentation contour to probable shapes. An implicit segmentation scheme is chosen instead of an explicit one, which is beneficial regarding topological flexibility and implementational issues. The mathematically sound probabilistic shape model enables the challenging integration of an explicit shape prior into an implicit segmentation scheme in an elegant formulation. A maximum a posteriori estimation is developed of a level set function whose zero level set best separates the organ from the background under a shape constraint introduced by the model. This leads to an energy functional which is minimized with respect to the level set using an Euler-Lagrangian equation. Since both the model and the implicitly defined contour are well suited to represent multi-object shapes, an extension of the algorithm to multi-object segmentation is developed which is integrated into the same probabilistic framework. The novel method is evaluated on kidney and hipjoint segmentation.

Zusammenfassung

Ein probabilistisches Framework für punktbasierte Formmodellierung in der medizinischen Bildanalyse

Die vorliegende Doktorarbeit konzentriert sich auf die Entwicklung eines auf Korrespondenzwahrscheinlichkeiten beruhenden punktbasierten statistischen Formmodells in einem mathematisch fundierten und geschlossenen Framework. Ein weiterer Schwerpunkt liegt in der Integration des entwickelten Modells in eine Segmentierungsmethode. Hier wird ein neuartiger Ansatz vorgestellt, in welchem explizit definiertes Formwissen mit einer implizit definierten Segmentierungskontur kombiniert wird.

In der medizinischen Bildanalyse gilt der Begriff der Form als wichtiges Merkmal für die Erkennung und die Analyse anatomischer Stukturen. Die Formmodellierung mittels des Konzeptes der statistischen Formmodelle verkörpert ein mächtiges Werkzeug, welches zu Lösungen für Analyse-, Segmentierungs- und Rekonstruktionsprobleme beiträgt. Ein statistisches Formmodell versucht, einen Satz von segmentierten Formbeobachtungen eines gegebenen Organs optimal durch eine mittlere Form und ein Variabilitätsmodell zu repräsentieren. Eine große Herausforderung für jeglichen statistischen Ansatz stellt hierbei die Bestimmung von Korrespondenzen zwischen den Formen dar. Die übliche Annahme von 1-zu-1 Korrespondenzen ist problematisch aufgrund der Unsicherheiten die Genauigkeit der Segmentierung betreffend als auch aufgrund der allgemeinen Schwierigkeit, exakte Korrespondenzen zu lokalisieren.

In dieser Arbeit wird als Lösung für das Punkt-Korrespondenzproblem eine Korrespondenzwahrscheinlichkeit für alle Punktepaare berechnet. Dies bedeutet, daß die Formoberflächen durch Gauß'sche Mischverteilungen repräsentiert werden. Diese Herangehensweise erlaubt eine Formulierung der Modellberechnung in einem generativen Rahmen, in dem die Beobachtungen als zufällig durch das Modell generierte Stichproben interpretiert werden. Darauf aufbauend wird die Modellberechnung als Optimierungsproblem behandelt. Es wird ein Algorithmus zur Berechnung der Modell- und Beobachtungsparameter in einem einzigen Maximum-A-Posteriori Kriterium vorgeschlagen. Dies führt zu einem mathematisch fundierten und geschlossenen Framework.

Die Methode wird in einer Experimentserie an synthetischen und realen Daten evaluiert und validiert. Dafür werden adäquate Leistungsmaße und Metriken definiert, anhand derer die Qualität des neuen Modells mit den Qualitäten eines klassischen punktbasierten Modells und eines etablierten oberflächenbasierten Modells, die beide auf 1-zu-1 Korrespondenzen beruhen, verglichen wird.

Es wird ein Segmentierungsalgorithmus entwickelt, welcher das im Modell enthaltene Vorwissen über die Formen einsetzt, um die Segmentierungskontur auf wahrscheinliche Formen zu beschränken. Statt eines expliziten wird ein impliziter Segmentierungsansatz gewählt, da dieser in Bezug auf topologische Flexibilität und Implementierungsfragen Vorteile aufweist. Das mathematisch fundierte probabilistische Formmodell ermöglicht auf elegante Weise die anspruchsvolle Integrierung von explizit repräsentiertem Vorwissen über die Form in einen impliziten Segmentierungansatz. Es wird eine Maximum-A-Posteriori Schätzung einer Levelsetfunktion so formuliert, daß das zugehörige Zero-Levelset das zu segmentierende Organ unter Einbeziehung der Formbeschränkung, die durch das Modell gegeben ist, optimal vom Hintergrund trennt. Dies führt zu einem Energiefunktional, welches unter Nutzung der Euler-Lagrange-Gleichung in Richtung der Levelsetfunktion differenziert wird. Da sowohl das Modell als auch der Segmentierungsansatz gut geeignet sind für die Beschreibung von Formen, die aus mehreren Objekten bestehen, wird eine Erweiterung des Algorithmus zu einer Multi-Objekt-Segmentierung entwickelt und in die gleiche probabilistische Formulierung integriert. Der Segmentierungalgorithmus wird an Nierendaten und Hüftgelenkdaten evaluiert.

Contents

List of Figures

List of Tables

CHAPTER 1

Introduction

1.1 Motivation

Since the discovery of X-rays in 1895, many different imaging techniques have been developed which gain visual access to the interior of a closed body without opening it. Nowadays, these techniques are widely used in health-care and biomedical research and constitute a substantial part of the clinical practice. In order to facilitate the interpretation of the generated body images, a multitude of medical image analysing methods has been realized which support the physicians in the fields of diagnostics, surgical planning and image guided surgery as well as medical research. With the progress of image acquisition techniques, the modeling of anatomical structures in 3D or even 4D has become an important component in medical image computing as these models offer an additional perspective for the surgeons and are used for model-based analysis, segmentation and classification problems. A popular approach for shape modeling is constituted by statistical methods which aim to represent an organ by statistical shape models. As opposed to a single 3D model or an atlas of an organ which are only (typical) shape examples, a statistical shape model represents a set containing segmented organs by a mean shape and a variability model. Hence, statistical shape models incorporate a priori shape knowledge drawn from many organ examples. Especially for segmentation problems, the application of statistical shape models has been proven to be very successful for a wide range of anatomical structures in CT, MR and ultrasound images.

The idea of doing statistics on shapes first leads to the problem of distinctly defining the concept of a shape. A well known definition proposed by the mathematician D. G. Kendall in 1984 reads as follows: "Shape is all the geometrical information that remains when location, scale and rotational effects are filtered out from an object" [Kendall 1984]. However, when dealing with anatomical structures, a more flexible definition is needed which also recognizes *deformable* objects based on their shapes. Therefore, at least effects like flexion and shearing have to be integrated. This means that the shape analysis methods are applied only after an affine alignment of the respective deformable objects.

The characteristics of a statistical shape model essentially depend on the representation of the shape surface. Basically, a surface can be seen as a boundary which separates geometrical regions in 3D. Mostly, it is represented explicitly using meshes or point clouds or implicitly based on distance functions. In order to compute a surface representation for a binary object, a sampling of the isosurface has to be performed. The sampling is a crucial step which - together with the imaging technique - determines the detailedness of the resulting surface model.

A fundamental problem for the computation of statistical shape models is the determination of correspondences between the observations. In order to quantitatively analyse shape differences, a method is needed to locate a corresponding point location on one shape for a given point location on another shape. Obviously, the solution to this problem always depends on the shape representation. Most current methods rely on surface-based representations and work with one-to-one correspondences. They do not consider the uncertainties neither of the segmentations nor of the sampling output nor of the registration results. Moreover, even for the utopian case of perfect segmentation and continuous surface representation, correspondence determination is never non-ambiguous but for reproducible prominent landmark locations.

The motivation of this thesis is to develop an alternative statistical shape model which takes into account the uncertainties of the whole scene and to investigate methods of applying this model for automatic segmentation. Most current algorithms compute the mean shape and variability model on a step-by-step basis. Therefore, a specific goal of this thesis is to realize the model computation in a sound mathematical framework.

1.2 Objectives

Following the motivation phrased in the previous section, we argue that when segmenting anatomical structures in noisy image data, the sampled surface points only represent probable surface locations and not necessarily the exact "true" shape surface. Besides, the choice of sampling method significantly influences the statistical analysis of the shapes. For instance, when the same binary object is sampled twice with different resolutions, the resulting surface representations will not be identical which makes the determination of exact correspondences impossible. Moreover, even for theoretically perfectly continuous surfaces, a unique and reproducible determination of correspondences is an open problem. It even becomes impossible if one of the surfaces features a shape detail that the other one lacks. For an illustration, imagine a reconstructed head of the sphinx containing a nose, and then imagine the challenge of determining a corresponding point for the tip of that nose on the original sphinx head. It is desirable to explicitly model the uncertainties of the scene. In order to come up with a realistic modeling of a surface based on the sampled points, the goal is to investigate the possibilities of representing the shapes in a probabilistic framework where each sampled surface point is drawn from a 3D probability density function (typically a Gaussian).

Most algorithms in the state-of-the-art approach the problem of model computation based on a set of segmented organ shapes for which the best statistical shape model must be computed. In order to develop a theoretical foundation of the algorithm it might be of interest to adopt an alternative view on the problem of model computation. The focus of this thesis lies on the development of a statistical shape model based on correspondence probabilities in a sound mathematical framework and its application in medical image segmentation.

These demands lead mainly to the following three objectives:

- **Development of a probabilistic framework to compute a generative statistical shape model based on correspondence probabilities:** The first problem tackled is the computation of a generative statistical shape model that optimally represents the shapes in a training data set. The aim is to design a point-based parametric model which allows the modeling of variability for each point. This might help physicians to physically interpret the deformations. The focus lies on the development of a generative probabilistic framework which includes all variables needed to describe the scene. Additionally, the framework has to integrate a solution to the correspondence problem.

- **Development of a deformable model segmentation in a probabilistic framework:** A major problem in medical image processing is the automatic segmentation of anatomical structures. Therefore, the second problem to be dealt with is the integration of the generative statistical shape model into an automatic segmentation scheme. The objective is to develop a sound mathematical formulation which is based on the same probabilistic assumptions as the framework for the computation of the statistical shape model. It is intended to develop a segmentation algorithm which enables the segmentation of objects with non-spherical topology as well as multiple-object shapes.

- **Evaluation and validation with respect to existing methods:** A main advantage of working with point-based shape representation is the simplicity of the resulting model with respect to its power. On the other hand, surface-based models generally feature better quality measures than point-based models. However, the quality of the surface information they use depends on image quality and on the segmentation method (which is often based on points drawn by experts). In order to place the new method in the state-of-the-art, it is crucial to evaluate the quality of the probabilistic model in comparison with other statistical shape models, investigate applications like classification methods and expose advantages and limits of the new model. Secondly, an evaluation of the segmentation method on different real data segmentation problems is needed in order to identify the strengths of the method with respect to the state-of-the-art.

1.3 Structure of Manuscript

This thesis is organized pursuing these motivation and objectives as follows: Chapter 2 provides information about the state-of-the-art in statistical shape analysis. Chapter 3, 4 and 5 contain the main contributions regarding the development and application of a new statistical shape model and a new level set segmentation method relying on the model. Chapter 6 concludes the manuscript. In the following, a condensed summary is given for each chapter.

In **Chapter 2** the background information needed about current methods in statistical shape analysis is summarized. It begins with a description of the state-of-the-art regarding the use and types of statistical shape models. Then the point correspondence problem is covered in detail before different methods for the computation of statistical shape models and their applications are presented.

In **Chapter 3** an approach to the problem of designing a generative statistical shape model is developed [Hufnagel 2007b, Hufnagel 2008b]. First, a solution to the point correspondence problem is derived by representing the shapes by Mixtures of Gaussians. Following that, a sound and unified framework is developed for the computation of model parameters and observation parameters as well as nuisance parameters, and a maximum a posteriori estimation is formulated which leads to a global criterion. Explicit formulas are derived for its optimization with respect to all parameters. Finally, practical aspects of the implementation and adaptions of the algorithm for special cases are discussed.

In **Chapter 4** an evaluation and validation of the generative Gaussian Mixture statistical shape model as developed in this thesis is performed. First, the choice of performance measures is established. Then, the performance of the new statistical shape model is compared to the performance of a classical point-based statistical shape model based on the iterative closest points registration and the principal component analysis [Hufnagel 2009a]. Furthermore, the performance of the new statistical shape model in comparison with a surface-based statistical shape model which is computed by the minimum-description-length approach is evaluated. The evaluation is done on synthetic and real data. Different examples covering convex and non-convex as well as spheric and non-spheric shape data are chosen.

In **Chapter 5** an automatic segmentation algorithm is developed which employs the a priori shape knowledge inherent in the new statistical shape model. After explaining the benefits of employing a non-parametric segmentation contour instead of a parametric one, the problem of integrating an explicitly represented statistical shape model into an implicit segmentation scheme is tackled. To our knowledge, very few works considered that option. The problem is solved by developing a novel maximum a posteriori estimation of the segmentation contour which is optimized based on the image information as well as on the statistical shape model information. Here, the respective steps which finally lead to a sound probabilistic segmentation scheme are explained elaborately. It is demonstrated in detail how to optimally exploit the image information to guide the evolution of the contour, and the implemented techniques to determine an initial positioning of the segmentation contour are presented. As the model is based on correspondence probabilities instead of one-to-one correspondences, the modeling and segmentation of non-spheric and multi-object structures is possible. Consequently, an extension of the algorithm to multi-object segmentation is developed which is integrated in the same framework by adapting the correspondence criterion. Experiments are designed and conducted in order to validate the segmentation method on kidney data and on hip joint data. Finally, the results are critically discussed, and the advantages and limits of this segmentation method are revealed. Part of this chapter is published in [Hufnagel 2009c].

In **Chapter 6** the contributions of this thesis are discussed and perspectives for future work are given.

Appendix A contains the mathematical background and detailed explanations for some of the derivations in the manuscript.

CHAPTER 2

Current Methods in Statistical Shape Analysis

The extraction of information out of 2D or 3D images often relies on the detection, recognition and interpretation of shapes and shape variabilities. This directly involves the (mathematical) representation of shapes as well as methods to account for and measure the morphological differences. Even though in clinical routine shape analysis is frequently done by viewing the images alone, there is a wide range of applications where automatical methods with formalized metrics are needed for overall data interpretation and shape statistics. This chapter is dedicated to the description of these methods and is divided as follows: First, the importance of shape modeling in medical image analysis is outlined and the concept of statistical shape models and their representations are discussed in section 2.1. Following that, we expand on the fundamental problem of determining correspondences between shapes and on several methods of solution (section 2.2) which directly leads us to discuss the associated statistical shape models in section 2.3. Section 2.4 explores the benefits of statistical shape models for medical image segmentation and discusses explicitly and implicitly represented shape priors.

2.1 Shape Modeling in Medical Imaging

Shape models are used for a wide range of medical imaging problems like segmentation, reconstruction or shape analysis. In this section, a condensed overview about the domain of shape analysis techniques in nowadays medical research is given (section 2.1.1) and then the subject of doing statistics on different shape representations is introduced (section 2.1.2).

2.1.1 Shape Analysis

The thorough analysis of organ morphology is driven by the hope of better understanding organ shape characteristics and how diseases might affect them. The idea is to find information based on the shape deformation or shape differences which eventually help in the diagnostics, especially in the neuroimaging community. The modeling of shape and the measuring of morphological changes in shape instances is also of great interest for the discrimination between healthy and pathological anatomical structures. An intuitive approach for detecting shape difference is the measurement of the global shape volume, however, this feature is often not significant with respect to the studied disease. This has been shown for example by Gerig et al. [Gerig 2001] based on the detection of group differences in hippocampal shapes in schizophrenia. Results of higher significance are often obtained by performing a local shape analysis. A wide range of approaches exists in the literature which can be roughly categorized according to the (shape) features chosen to do the statistics on. In the following, an overview of developments in that field is given by means of exemplarily selected publications.

Early methods proposed to analyse and compare the *transformation fields* obtained when registering an organ to a template, which is used e.g. in the work of Davatzikos et al. [Davatzikos 1996] who analyse the morphology of the corpus callosum. A similar idea is applied in the work of Boisvert et al. [Boisvert 2008] who model spine shape deformation by a vector of rigid transformations. First efforts in mathematically capturing morphology changes by doing statistics on anatomical *landmarks* were undertaken by F.L. Bookstein [Bookstein 1986, Bookstein 1991]. The concept of statistical shape analysis based on landmarks and pseudo-landmarks was taken on by Dryden and Mardia [Dryden 1993] for the detection of gender related differences in monkey crania and by Bookstein [Bookstein 1997] for the detection of brain differences in schizophrenia patients. In both approaches, the shape variations are measured based on Procrustes or Riemannian distances. Another shape analysis method is based on a *medial shape description* to model local and global changes as e.g. used by Styner et al. [Styner 2003b] who analyse the hippocampus shape of schizophrenia patients. In several works the shapes are represented by *distance functions* whose feature vectors are used as input for a learning algorithm, e.g. in the work of Golland et al. [Golland 2001] who compute a classifier for healthy and pathological hippocampal shapes in schizophrenia or in the work of Kodipaka et al. [Kodipaka 2007] whose Kernel Fisher discriminant distinguishes between controls and epileptics by analysing the shape of the temporal lobe or in the work of Tsai et al. [Tsai 2005] who propose an EM formulation to automatically label lung shapes represented by level set functions to belong to the normal or the emphysema shape class. In the work of Peter et al. [Peter 2006a], shapes are represented by a Gaussian Mixture Model on the landmarks, and the shape differences (here of corpus callosum shapes) are measured using geodesic distances under the Fisher-Rao metric.

Naturally, all of these approaches have their strengths and weaknesses. The choice of feature suited as input for the statistical analysis depends on the representation of the shapes as well as on the demands of the application. The work done in the framework of this thesis concentrates on the category of shape analysis based on point representations since statistics on points are easily interpretable and have a physical significance. The general concept however is not necessarily confined to that category.

2.1.2 Doing Statistics on Shapes

Commonly, a shape class can be described by one typical shape example of the respective organ. However, this approach is neither specific nor mathematically accurate. In order to reliably describe a shape class, we need to statistically evaluate the shapes of as many observations of the organ as possible. This is usually done in four steps: First, a training data set which contains segmented observations of the respective organ has to be provided. Next, the observations have to be aligned in a common reference frame in order to eliminate pose variations. Then, a mean shape which optimally represents all aligned observations can be computed. Finally, a variability model accounting for the shape differences is determined. The variability contains information about how much and in which way the mean shape can be deformed while still representing a plausible anatomical structure.

In the state-of-the-art, shape models containing a mean shape and a variability model are referred to as statistical shape models (SSMs). The methods implementing the alignment as well as the statistical methods used for the computation of mean shape and variability model depend on the representation of the observations. An intuitive and widely-used method is to compute SSMs on observations represented by (triangulated) points which are distributed over the surface of the shapes. These so-called point distribution models (PDMs) are either based on anatomical landmarks [Huysmans 2005], on pseudo-landmarks that are strategically distributed over the observations' surfaces (e.g. [Frangi 2001, Rajamani 2004]) or on points reconstructed from implicit surfaces (e.g. [Kohlberger 2009]) or on a combination of these. Point-based shape samples represented by a number of N points in 3D are usually described by a shape vector $S_k \in \mathbb{R}^{3 \times N}$ containing the point coordinates. The alignment to a common reference frame is often performed by a mesh-to-mesh registration over the shape vectors. The statistic evaluation then uses the aligned shape vectors as input for computation of mean shape and variability model.

For these steps, a notion of correspondence has to be defined. A common approach is to assume and determine one-to-one point correspondences over all observations. In that case, the coordinates of corresponding points are sorted in corresponding entry positions in the shape vectors so that for all shape pairs S_k and S_l the i-th element $S_k(i)$ corresponds to $S_l(i)$ for all $i = 1, ..., 3N$. The computation of the mean shape is then straight forward with $\bar{M} = \frac{1}{n} \sum_{k=1}^{n} S_k$ for a number of n observations. The subsequent computation of variation modes is usually accomplished by a principal component analysis (PCA) on all observations and the mean shape. The variation modes $\in \mathbb{R}^{3N}$ are pairwise orthogonal and span the shape space of the SSM. Mathematically, the representation of a random shape M in the shape space spanned by the variation modes can be formulated using a linear model:

$$M = \bar{M} + Pb$$

where the matrix $R \in \mathbb{R}^{N \times N'}$ with $0 < N' \leq N$ contains the variation modes in its rows and the vector $b \in \mathbb{R}^N$ contains the coefficients which control the extent of deformation. The variation modes are ranked according to their variance. For the usage of an SSM, commonly only the largest modes of variation are taken into account.

The employment of the PCA is not confined to point representations but can be employed to other applications where the shape properties are encoded in a feature vector. Early methods include the representation of shapes by spherical harmonics (SPHARM) which parameterize the surface by a mapping on the unit sphere [Brechbühler 1995, Székely 1996] or by Fourier surfaces which employ an elliptic Fourier decomposition of the boundary and use the Fourier coefficients as feature vectors [Staib 1996, Floreby 1998]. The statistics are thus done in parameter space. Recently, the representation of SSMs in implicit frameworks has become of interest as level set based segmentation is explored more deeply. Here, the observations in the training data set are often represented by signed distance maps. The alignment of the observations and the subsequent statistics are then done directly on the distance maps which are used as feature vectors with individual voxels being the vector components. The variability models can simply be computed by a principal component analysis [Leventon 2000a] or by using more challenging methods which for example account for local variations as well [Rousson 2002]. Another strategy represents the surfaces by medial models which consist of a centerline and vectors stretching from there to the organ surface [Pizer 1999, Styner 2001]. Here, correspondence between shapes are defined on the medial manifold. For computing the variability of manifold-valued data, a principal geodesic analysis is introduced which is a direct generalization of principal component analysis.

It has to be kept in mind that the PCA is done under the assumption that the shape vectors are samples of a random variable under a normal distribution. This is only the case if the shape differences in the training data set are normally distributed which is difficult to establish. Another theoretical problem occurs as the dimensions of the shape representation nearly always exceed the number of availabe samples. Besides, the PCA is optimal in a least-square sense and therefore sensitive to outliers and lastly, all shapes have to be represented by feature vectors of equal lengths. Nevertheless, the employment of the PCA for SSM computation has been proven to come to acceptable results and is successfully applied in practice. An alternative for non-normally distributed data is offered by the so-called independent component analysis (ICA) [Hyvärinen 2001]. The ICA decorrelates the components by maximizing their statistical independence. Another interesting approach is to do a principal factor analysis (PFA) which leads to variation modes that are more easily interpretable in medical sense [Ballester 2005, Reyes 2009]. However, these methods have the disadvantage that the variation modes cannot be ranked easily which poses a problem for dimensionality reduction.

2.2 The Correspondence Problem

A fundamental problem when computing statistical shape models is the determination of correspondences between the observations in the training data set. Mathematically, this problem does not have a unique solution and depends heavily on the definition of 'shape' as well as on its representation. For shapes represented as contours in 2D, usually landmarks are determined manually by first choosing exposed features as landmarks, for example the fingertips of a hand as well as the points between the fingers, and by then adding a fixed number of equidistant landmarks between these. In that way, the correspondences from one labeled shape to the next equally labeled one is straightforward and uniquely defined. In 3D, however, a manual determination of correspondence is hardly feasible as it is very time-consuming in general. In particular, the pinpointing of exact correspondences without relying on clear anatomical landmarks on 3D surfaces is an impossible task. In order to automatize the detection of landmarks, several methods extract shape features such as high surface curvatures (e.g. [Benayoun 1994]). Mostly however, automatic determination of correspondences is done by performing a registration of model and observation. Obviously, the solutions to the correspondence problem highly depend on the shape representations. For meshes, a straightforward approach is to compute a similarity transformation found by least-square point distance minimizers. For non-linear registration, often spline-based deformations are used. Another approach is the matching of an atlas or template mesh to all observations in the training data set. The warped meshes have to be relaxed in order to fit the observations. This can be done for example by using a Markov random field regularization as proposed by Paulsen and Hilger [Paulsen 2003] or by employing a spring-mass model based on the surface point set and the connecting edges as realized by Lorenz and Krahnstöver [Lorenz 2000]. A method for volumetric representations is to compute a volumetric atlas with manually added surface landmarks and then register the atlas to volumetrically represented observations. The warped landmarks then determine the correspondences.

In this section, two popular methods for correspondence determinations are described based on different shape representations which play a role in the remainder of this thesis: First, the classical Iterative Closest Points (ICP) registration algorithm that finds one-to-one correspondences between two unstructured point sets is explained. Then, an alternative approach to correspondence determination using spherical harmonics surfaces parameterization is presented. Here, the correspondences are computed by a registration between the parameterizations of the shapes. As an example for methods which solve the correspondence problem in a groupwise optimization approach together with the SSM computation the maximum description length (MDL) approach is summarized in section 2.3. A comprehensive comparison of different solutions to the correspondence problem can be found in [Styner 2003c].

2.2.1 Iterative Closest Point Algorithm

The Iterative Closest Point algorithm is an efficient method used for registration of 2D and 3D shapes as first shown and elaborately explained 1992 in [Besl 1992]. The ICP registration is an interesting approach as it can be used for different representations of geometric data like point sets, triangle sets, and implicit or explicit surfaces. It is applied to registration problems where the point correspondences are not known in advance. The ICP algorithm offers many recognized advantages as it does not need preprocessing or local feature extractions in normal applications, it is suited for parallel architectures and it can handle average noise. Following, a simple definition of the ICP algorithm and its application to point cloud registration is given.

Let S be a set of N_s points s_i which describe the observation and M be a set of N_m points m_j which describe the model. The ICP algorithm will match each observation point s_i with one of the model points. Based on those matches, a transformation T is sought which registers the observation with the model. The closest point operator CP is defined as a distance metric

$$CP(s_i, M) = \min_{m_j \in M} \|m_j - s_i\|.$$

We use $m_j^i = CP(s_i, M)$ where m_j^i is the closest point in M to a given scene point s_i. The ICP algorithm computing T is implemented as follows:

1. $T^{(0)} = T^k$ is chosen as initial estimate of the transformation T.

2. Repeat for k iterations or until convergence:

   - Compute the closest point $m_j^i \in M$ in the model for each observation point $s_i \in S$. The collection of resulting point pairs (s_i, m_j^i) is called *set of correspondences* C with

     $$C_{k-1} = \cup_{i=1}^{N_s}\{s_i, CP(T^{k-1} \star s_i, M)\}.$$

   - Compute T^k that minimizes the mean square error between all point pairs in C.

For a rigid registration, the application of T to S looks like this

$$T \star s_i = Rs_i + t \qquad \forall i$$

with the rotation matrix $R \in \mathbb{R}^{3x3}$ and the translation vector $t \in \mathbb{R}^3$. The minimization of the error between all point pairs in C is computed using the Least Squares criterion:

$$\begin{aligned}
T &= \operatorname*{argmin}_{T} \frac{1}{N_s} \sum_{i=1}^{N_s} \|m_j^i - T \star s_i\|^2 \\
&= \operatorname*{argmin}_{R,t} \frac{1}{N_s} \sum_{i=1}^{N_s} \|m_j^i - Rs_i - t\|^2.
\end{aligned}$$

The ICP algorithm converges always monotonically to the nearest local minimum where "nearest" is meant in the sense of a mean-square distance metric.

As main disadvantage it must be noted that the ICP is susceptible to gross statistical outliers. Several approaches deal with this problem by e.g. proposing robust estimators [Zhang 1994, Masuda 1996]. Moreover, as any method minimizing a non-convex cost function, the ICP lacks robustness with respect to the initial transformation because of

Figure 2.1: *A correspondence problem: One shape features two bumps, the other only one. How can we determine correspondences between the two?*

local minima. This problem has been tackled by the work of Rangarajan et al. who use multiple weighted matches based on Gaussian weight [Rangarajan 1997b] and based on Mutual Information [Rangarajan 1999].

Overall, the ICP algorithm and its derivatives work well for a lot of registration problems. However, the determination of one-to-one correspondences between unstructured point sets is difficult when e.g. one shape features a certain structure detail and the other one does not, see figure 2.1. Moreover, in the absence of (anatomical) landmarks, the determination of correspondence depends heavily on the sampling of the shape. To overcome this problem, the Expectation Maximization - Iterative Closest Points (EM-ICP) algorithm introduces *correspondence probabilities* instead of exact correspondences. This concept is explored in section 3.2.

2.2.2   Spherical Harmonic Description

The use of spherical harmonics for statistical shape modeling was introduced by Brechbühler et al. in 1995 [Brechbühler 1995] in order to approximate one-to-one corresponding points on different entities containing inclusions and protrusions. As opposed to the use of a torus parameter space using Fourier descriptors as proposed in [Staib 1992], the SPHARM surface description maps the observation surfaces into a spherical two-coordinate space, so it can only be considered for shapes with spherical topology which applies for most anatomical structures. If the mapping includes an optimization of the distribution of nodes on the sphere, correspondences can then be established directly by the parametric description.

Surface objects with spherical topology can be parameterized by two polar variables, the longitude $\theta = [0, ..., 2\pi]$ and the latitude $\phi = [0, ..., \pi]$. Two vertices have to be selected as the poles for this process. The latitude should grow smoothly from 0 at the north pole to π at the south pole. The longitude on the other hand is a cyclic parameter. Let x, y and z denote Cartesian object space coordinates. The function which specifies the mapping of the coordinates from the unit sphere on the surface is specified with

$$v(\theta, \phi) = \begin{pmatrix} x(\theta, \phi) \\ y(\theta, \phi) \\ z(\theta, \phi) \end{pmatrix}.$$

where $v(\phi, \theta)$ runs over the whole surface object.

These coordinate functions could be parameterized by various basis functions as e.g. B-splines or wavelets. The SPHARM algorithm makes use of spherical harmonics as they offer the advantage of hierarchical shape representation which finally facilitates the correspondence determination [Brechbühler 1995]. Typically, the following truncated series expansion is used:

$$v(\theta, \phi) = \sum_{r=0}^{R} \sum_{-r}^{r} c_r^m Y_r^m(\theta, \phi)$$

where Y_r^m denotes the function of degree r and order m with $Y_r^m : [0, 2\pi] \times [0, \pi] \to \mathbb{C}$. A complete definition can be found in e.g. [Bronstein 2000]. The shape descriptor coefficients c_r^m are 3D vectors with components (x, y, z). Formally, the coefficients are computed by the inner product of function v and the basis function

$$c_r^m = \int_0^\pi \int_0^{2\pi} v(\theta, \phi) Y_r^m(\theta, \phi) d\phi \sin \theta d\theta. \tag{2.1}$$

Eventually, each shape surface S_k is uniquely described by a set of descriptor coefficients $C_k = c_{k,r}^m$.

Due to the hierarchical shape representation, in practice the level of shape details which are modeled depends on the maximal degree R in the spherical harmonics. The parameterization for degree 1 maps the surface to an ellipsoid. In order to determine shape point correspondences by parameterization to a sphere, the mapping between surface and sphere must be bijective which is described in this case by

$$\begin{pmatrix} x \\ y \\ z \end{pmatrix} = \begin{pmatrix} \sin \theta \cos \phi \\ \sin \theta \sin \phi \\ \cos \theta \end{pmatrix}.$$

Furthermore it must be continuous so that neighbouring points on the shape surface are mapped to neighbouring locations on the sphere. The mapping function should be topology-preserving, and distortions which inevitably appear when mapping a surface facet to a spherical square should be minimal. This is done by solving the surface parameterization as a constrained optimization problem with respect to the optimal coordinates for all surface points [Brechbühler 1995]. Another problem occurs as the coefficients obtained by approximating equation (2.1) depend on the rotation of the surface in space. Thus, for the determination of correspondences between different shape observations, a rotation of all observations to a canonical position in parameter space is needed. This can be done using the spherical harmonics of degree 1 by rotating the parameter space so that the north pole (where $\theta = 0$) is positioned at one end of the shortest main axis of the ellipsoid, and the point where the Greenwich meridian ($\phi = 0$) crosses the equator (where $\theta = \pi/2$) is positioned at one end of the longest main axis.

The statistics on the shapes are now done by evaluation of the shape descriptors. The mean shape then is described by the spherical harmonics using the set of averaged shape descriptor coefficients $\bar{C} = \frac{1}{N} \sum_k^N C_k$ and the principal component analysis is done using the covariance matrix $\frac{1}{N-1} \sum_k (C_k - \bar{C})(C_k - \bar{C})^T$. A point distribution model can than be generated directly by linear mapping [Kelemen 1999].

While the SPHARM parameterization is capable to smoothly represent high levels of shape details, it suffers from the fact that for shapes featuring rotational symmetry in the spherical harmonics of degree 1 the mapping to the canonical position in parameter space is not unique. Therefore, the correspondence determination for such shapes becomes inappropriate as shown in a study on e.g. femoral heads by Styner et al. [Styner 2003c].

2.3 Computation of Statistical Shape Models

In order to compute a SSM, a sufficiently large training data set with segmented organ observations is needed. Obviously, the training data set should only contain shapes conforming to the shape class which is modeled, that is, for a SSM of normal organ variability, only healthy patient data is permitted. Each observation has to be segmented accurately. This is mostly done manually or semi-automatically by medical experts who delineate the organ contours slice by slice in medical images. Some organs can be segmented also in 3D under the support of automatic techniques like volume growing of thresholding. For binary segmentation, the conversion to a surface representation is typically performed by the Marching Cubes algorithm [Lorensen 1987]. The first step is commonly the alignment of the observation in a reference coordinate system. Then, a mean shape and a variability model are computed such as to optimally represent the shapes in the training data set. Here, the accurate detection of correspondence between the shapes plays an important role regarding the quality of the final SSM. The resulting SSM produces new plausible shapes or represents unknown shape observations of the same organ in different patients or under different conditions.

In this chapter, the computation of two widely-used point distribution models is summarized: Section 2.3.1 describes the classical Active Shape Models (ASM) while section 2.3.2 presents a method to build ASMs using gradient descent optimization of the maximum description length.

2.3.1 Active Shape Models

With the introduction of the 'Active Contour Models' (ASMs) or 'Snakes' in 1988 by
Kass et al. first attempts were made to integrate a priori knowledge into the segmentation process by forcing the segmentation contour to comply to a certain amount of
smoothness [Kass 1988]. The technique makes use of an iterative energy minimization where only local shape constraints are applied. Cootes et al. adopted an iterative
approach but instead of applying a simple snake contour, they developed a point distribution model or 'Active Shape Model' to incorporate a priori knowledge about the
shape [Cootes 1992, Cootes 1995]. When applying the ASM to segmentation, they use
global shape constraints.

Let us describe the N observations S_k in the training data set by meshes consisting
of n_k points $s_{ki} \in \mathbb{R}^3$. Furthermore, let us assume that $n_k = n \; \forall k$ and that the
points with the same index i correspond. The set of observations can then be aligned
by translation, rotation and anisotropic scaling so that the least squared differences
between all corresponding points is minimized. This is done by an affine transformation
T_k. For an example see figure 2.2(a). If the alignment is omitted, the variation in size
and pose are included in the final variability model. The points $\bar{m}_i$ of the mean shape
$\bar{M}$ are then computed by averaging over all aligned corresponding observation points
$\bar{m}_i = \frac{1}{N} \sum_{k=1}^{N} T_k \star s_{ki}$. For an illustration see figure 2.2(b). In order to compute the
variability model, a principal components analysis (PCA) is performed. Under the
assumption of dealing with normally distributed data samples, the PCA determines a
linear transformation which transforms the data into a coordinate system where the axes
(= eigenvectors) lie in the same direction as the greatest correlations in the data. By
transforming the data into the new coordinate system, the correlations of the original
data set become uncorrelated. Thus, the new axes lie in the directions of the greatest
variance of the transformed data set. Hence, the data is represented in a system where
its similarities and differences can be seen clearly which makes the PCA a well-suited
tool in the description of shape variability. The N actual eigenvectors v_p and associated
eigenvalues λ_p are computed by e.g. doing a diagonalisation on the covariance matrix
with elements $cov_{ij} = \frac{\sum_{k=1}^{N} (s_{ki} - \bar{m}_i)(s_{kj} - \bar{m}_j)^T}{N-1}$, so $v_p \in \mathbb{R}^{3n}$ which amounts to one 3D
eigenvector v_{ip} per mean shape point $\bar{m}_i$, see figure 2.2(c). A plausible new instance of
the shape class can now be modeled by

$$M = \bar{M} + \sum_{p=1}^{N} \omega_p v_p \qquad (2.2)$$

where $\omega_p \in \mathbb{R}$ are the deformation coefficients which are typically constrained to $\omega_p \leq$
$3\lambda_p$ in order to only generate plausible shapes. Furthermore, a shape analysis can be
done by interpreting the deformations according to the eigenmodes with the greatest
eigenvalue (see figure 2.2(d,e,f)).

In order to better adapt the ASM to segmentation, Cootes et al. proposed the
Active Appearance Models (AAMs) which incorporate a priori knowledge not only about

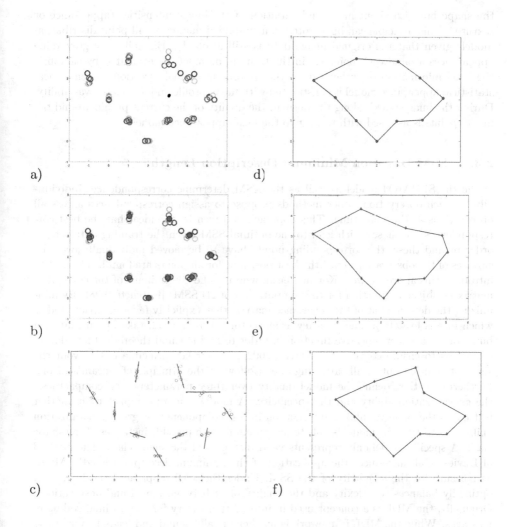

Figure 2.2: *ASM example. a) Aligned observations of a training data set. Each of the 5 observations is represented by 10 points in 2D and depicted in another colour. b) Mean shape point cloud depicted by filled circles. c) axes of first eigenmode depicted for each of the corresponding points. d) Mean shape $\bar{M}$ of point distribution model. e,f) Mean shape deformed according to first eigenmode $\bar{M} - 3\lambda v_1$ and $\bar{M} + 3\lambda v_1$.*

the shape but also about mean and variation of the image intensities (appearance or texture). This principle can be adapted in a simplified manner to all point distribution models given that the original image data is still available. Basically, the grey value appearances around each point s_{ki} in the training data set are evaluated by sampling the pixel information on either side of the contour in normal direction. Then a local statistical appearance model is constructed with mean profile and associated variability. During the image search along the normal, the quality of the current profile around the model points is assessed with respect to the local appearance model.

2.3.2   SSM Based on Minimum Description Length

While the SPHARM model as well as the ASM determine correspondences individually for each observation, other methods propose to assign correspondences across all observations at the same time. This approach is driven by the idea that the best correspondences are those which lead to the optimal SSM given the training data set. In order to find these, the corresponding points have to be moved individually over the surfaces of the observations until the best positions for all points are found. The first to introduce this approach were Kotcheff et al. who use the determinant of the covariance matrix as objective function for the computation of 2D SSMs [Kotcheff 1998]. By minimizing the determinant of the covariance matrix, they explicitly favor compact models which means low eigenvalues and few eigenvectors. Davies et al. take up on that idea but propose another objective function in order to find a sound theoretical foundation as well as to ensure convergence [Davies 2002c]. Their key principle is to favour the simplest solution out of all satisfying ones (following the principle of Occam's razor). Furthermore, they define the model quality over three parameters, the compactness, the generalization ability and the specificity. A model is more compact than another if it codes the same variability information in less components. A great generalization ability means that the model is able to describe unknown possible instances of the shape class. A specific model only represents valid instances of the shape class. The method of Davies et al. introduces the application of the minimum description length (MDL) as measure for the simplicity of the SSM. Under the MDL approach, the final SSM optimally balances complexity and the quality of fit between model and observations. Originally, the MDL is a concept used in information theory for the optimal coding of messages. While the MDL framework is mathematically sound and leads to very good results [Davies 2002a, Styner 2003b], the objective function is complex and computationally expensive. Several approaches have been proposed to reduce the complexity. Heimann et al. employ a simplified MDL cost function introduced in [Thodberg 2003] and use a gradient descent optimization to minimize it. They can show that their approach is faster and less likely to converge to local minima than previous approaches [Heimann 2005]. In this section, the principal concept of their algorithm is explained and the mesh parameterization as well as the optimal determination of correspondences used in their framework are outlined. The algorithm is constrained to SSMs of organs with spherical topology.

The cost function F which is based on the MDL of the resulting SSM is defined as

$$F = \sum_{p=1}^{n} L_p \quad \text{with} \quad L_p = \begin{cases} 1 + \log(\lambda_p/c_{cut}) & \text{for } \lambda_p \geq c_{cut} \\ \lambda_p/c_{cut} & \text{for } \lambda_p < c_{cut} \end{cases} \tag{2.3}$$

where λ_p denotes the squareroot of the eigenvalues of the covariance matrix. The parameter c_{cut} is a cutoff constant which describes the expected noise in the training data.

Regarding the mesh parameterization, a mapping of all surfaces to the unit sphere is performed. The mapping has to assign for every point on the surface of the mesh a unique position on the sphere. The problem of mesh parameterization is that of mapping a piecewise linear surface with a discrete representation onto a continuous spherical surface. In contrast to Davies et al. who use initial diffusion mapping, Heimann et al. create a conformal mapping that focuses on preserving angles. The function L maps each point s_i of the surface S to the unit sphere which results in a spherical parameterization of S. The mapping function is defined as $L : S \to \mathbb{R}^3$ with $|L(s_i)| = 1$ for all points s_i. The initialization is done by mapping each s_i to the position on the sphere corresponding to its normal vector. The optimal mapping is found by minimizing the string energy of the mesh as defined by Gu et al. who propose a variational method which can find a unique mapping between any two genus zero manifolds [Gu 2003]. Basically, two steps are executed: First, a barycentric mapping is performed which positions each point s_i at the center of its neighbouring points. Next, a conformal mapping is obtained by taking into account the angles between edges of the mesh for the parameterization. The mathematical proof of correctness of this approach is given in [Gotsman 2003].

After obtaining a conformal mapping L_k for each surface observation S_k, correspondences across the training data set are determined by mapping a set of spherical coordinates to each S_k. Subsequently, the optimal correspondences and therefore the optimal positions of all points on the surfaces have to be determined. To do so, Heimann et al. choose to modify the individual parameterizations L_k for all surfaces: In short, the corresponding landmarks of all observations are cleared of the mean and then stored in a matrix B'. By employing a singular value decomposition to $B = \frac{1}{\sqrt{n-1}}B'$, the eigenvectors and eigenvalues λ_p for the system of corresponding landmarks can be computed. This means that the λ_p in the cost function in equation (2.3) can be expressed in dependence of the singular values of B. Eventually, the cost function is minimized with respect to the elements of B by solving $\frac{\partial F}{\partial b_{ij}} = 0$. This derivation leads to a change for the individual landmark positions as shown in [Ericsson 2003] as it yields a 3D gradient for every landmark. In order to convert the gradients into optimal kernel movements $(\triangle\theta, \triangle\phi)$, $\frac{\partial F}{\partial(\triangle\theta, \triangle\phi)}$ is computed by

$$\frac{\partial F}{\partial(\triangle\theta, \triangle\phi)} = \frac{\partial F}{\partial b_{ij}} \frac{\partial b_{ij}}{\partial(\triangle\theta, \triangle\phi)}$$

where the surface gradients $\frac{\partial b_{ij}}{\partial(\triangle\theta,\triangle\phi)}$ are estimated by finite differences.

It has to be taken into account that when moving one landmark, the adjacent landmarks should be affected in a similar manner depending on their closeness. Therefore, a truncated Gaussian function is defined with

$$c(x,\sigma) = \begin{cases} \exp(\frac{-x^2}{2\sigma^2} - \frac{-(3\sigma)^2}{2\sigma^2}) & \text{for } x < 3\sigma \\ 0 & \text{for } x \geq 3\sigma \end{cases}$$

where x denotes the distance between the specific landmark and the center of the kernel and σ controls the size of the kernel. If a point at position x is moved by $(\triangle\theta, \triangle\phi)$, all other points are affected by $c(x,\sigma)(\triangle\theta, \triangle\phi)$. This re-parameterization is done iteratively over all landmarks and all observations. For a detailed derivation of this algorithm as well as an evaluation please refer to [Heimann 2005, Heimann 2007c].

Note that this approach only makes sense for mesh representations of surfaces but not for point cloud representations.

2.4   Segmentation Using Shape Priors

The goal of a segmentation process is the partitioning of an image into regions which are homogeneous regarding a certain number of characteristics. The multitude of image-based segmentation techniques can be roughly categorized into region-based, edge-based, and clustering methods. *Region-based methods* search for pixels amidst an area which fulfill a similarity criterion. A typical example are region-growing techniques which basically use a manually selected seed voxel and then automatically extract all voxels connected to the seed or connected to already extracted voxels featuring the same gray value [Haralick 1985]. Region-based methods are usually sensitive to noise and image-inhomogeneities. *Edge-based methods* detect contours which are defined by abrupt gray value changes in the image. For digital images, filtering masks (e.g. Prewitt, Sobel, Laplace) are used in order to compute the gradient images of first or second order. A disadvantage of edge-based methods is the fact that the resulting edges are often disconnected and consecutive boundary finding methods have to be employed. A widely-used clustering method is the thresholding segmentation which is a straightforward but often not very efficient technique where the pixels of an image are classified simply by determining if their gray value lies above or below an appointed threshold [Sahoo 1988]. The same idea applies to watershed approaches where the different gray levels are interpreted as topographic surfaces [Vincent 1991]. For multi-spectral image data, cluster-analysis methods are employed where the voxels are represented by feature vectors of higher dimensionality [Handels 2009]. Elaborate overview of these categories of segmentation techniques are given in [Gonzalez 2002].

Medical images tend to feature noise, contour gaps, intensity inhomogeneities and low contrasts. This is due to several problems: First, image acquisition systems yield relatively low signal to noise ratio. Secondly, soft tissue boundaries do not necessarily

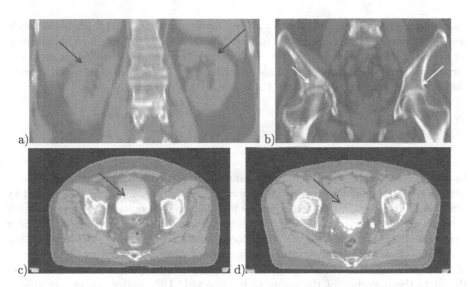

Figure 2.3: *Medical images. a) Kidneys in noisy CT data. b) Femur and hipbone CTs featuring contour gaps and low resolution. c),d) Bladder CTs featuring intensity inhomogeneities due to contrast agent and different filling levels.*

feature clear gradients (see figure 2.3(a)) and there is often a tissue variability in the same organ across patients (see figure 2.3(c,d)). Another problem are image artifacts due to patient motion or limited acquisition time which reduce the information content of the data (see figure 2.3(b)). Generally, methods which work on image information alone like region growing or thresholding or edge-filtering are sensitive to these characteristics. Furthermore, they are prone to errors under typical shortcomings of medical images like sampling artifacts and spatial alias effects. In order to robustify the segmentation process, an effective and popular approach is to employ models which incorporate a priori information about the structure to be segmented.

The concept of deformable models is explained in section 2.4.1, and the most important aspects of explicit and implicit shape priors are summarized in sections 2.4.2 and 2.4.3.

2.4.1 Deformable Models

A substantial part of segmentation methods nowadays is based on the concept of *deformable models* which was originally introduced for use in computer vision by Terzopoulos et al. [Terzopoulos 1986]. Since the work about Snakes (Active Contours) published in 1988 by Kass et al. [Kass 1988], deformable models are effectively used for segmentation, reconstructing, visualization and matching problems in 2D and 3D and have successfully been applied to a wide range of organs. A deformable model is usually represented by a contour or a surface. The deformation of the model is governed by means of energy minimization where the energy functional basically consists of one term which controls the resulting shape (internal energy) and one term which attracts the contour toward the boundary in the image (external energy):

$$E(C) = E_{int} + E_{ext}.$$

In a physical interpretation, deformable models are elastic bodies which respond in a natural way to the influence of external forces. The deforming forces are determined by image data like edges or textures as well as by smoothness conditions or a priori knowledge about the shape and location of the respective anatomical structures. The prior shape information renders the algorithm more robust and accurate [McInerney 1996]. A deformable model is usually initialized in an approximative manner around a region of interest. Then, it evolves from this initial rough solution to automatically improve the fit to the boundary of the region to be detected. Deformable models are able to model the complexity and sometimes significant variabilities of anatomical structures. For a thorough survey which focuses on the topological, geometrical and evolutional aspects of deformable models see [Montagnat 2001].

In the last years, the integration of a priori information about the shape has proven to be a very efficient approach which led to a multitude of robust automatic segmentation techniques for various medical applications. The key idea is to constrain the segmentation to plausible shapes. Mostly, statistical shape models (SSM) are employed. The different shape prior models can be divided into the following two main approaches: the parametric models which evolve corresponding the Eulerian formulation (section 2.4.2) and the implicit models which evolve corresponding to the Lagrangian formulation (section 2.4.3). In order to demonstrate the variety of segmentation methods which benefit from prior knowledge about the shape, a brief survey is given about some of the most popular applications: Explicitly represented SSMs have been successfully employed e.g. for pelvic bone segmentation [Seebass 2003, Lamecker 2004], for hipjoint segmentation [Kainmüller 2009] and for (scoliotic) vertebrae segmentation [Benameur 2003, Pekar 2001]. Furthermore, SSMs are frequently used for soft tissue segmentation as e.g. for liver segmentation from CT data [Lamecker 2003, Heimann 2007a] or for segmentation of aortic aneurysms from CT data [de Brujine 2002]. Other authors use implicit SSM for CT kidney segmentation [Tsaagan 2002]. Right from the start, SSMs were discovered to be beneficial in the segmentation of cardiac structures as the left ventricle [Staib 1996, Kaus 2004, Shang 2004] or the whole heart [Lötjönen 2004, Lorenz 2006]. Moreover, the use of SSMs is a widespread method in brain segmentation on MR images, e.g. by SPHARM modeling [Székely 1996], m-rep modeling [Pizer 2003] or explicit modeling [Zhao 2005a].

2.4.2 Explicitly Represented Shape Priors

With the presentation of the Active Shape Models (ASM) in 1992, Cootes and Taylor introduced a method to use explicitly represented point distribution models as shape priors for segmentation tasks [Cootes 1992]. The definition and mathematical formulations of such statistical shape models are given in section 2.3. In short, the segmentation techniques using the ASM method work as follows: First, the model is placed in the image. This initial placement favorably close to the structure to be segmented is often done manually. Next, for each model point a movement is suggested along its normal toward a position lying closer to the contour of the object to be segmented. Commonly, for each point a candidate contour position is determined by evaluating the neighbouring voxels in direction of the contour normal. The candidate quality of positions depends on boundary-based and/or region based features. For their appearance models, Cootes propose to use the normalized first derivatives of the profiles [Cootes 2001a]. Brejl et al. make use of a combination of grey values and grey value gradients [Brejl 2000]. Other appearance models include region-based features like the texture inside the shape [Cootes 2001b] or the creation of histograms on inside and outside regions [Broadhurst 2006]. Eventually, the optimal choice of appearance model depends on the image modality as well as the anatomical structure to be segmented as shown for example in [Heimann 2008]. After determining a candidate position for each point, the model is transformed and deformed to optimally approximate the candidate points. The deformation is constrained to lie in the model variability space. These updates of the model are iterated until the moving distance of model points falls under a certain threshold. A detailed explanation of the algorithm is given in [Cootes 2004].

The principal idea of ASM segmentation still forms the basis for numerous segmentation methods employing statistical shape models nowadays. However, the limits placed on the model parameters ensuring that the segmentation contour can only adapt to shapes which are probable regarding the underlying training data set are too constraining for many segmentation tasks. This is mainly due to the fact that the number of training observations is usually too small to represent all probable shape variabilities. To lighten the constraint, several authors proposed segmentation algorithms which balance between prior shape knowledge introduced by the SSM and image information. These algorithms range from using the converged SSM as initialization for additional refinement steps [Cootes 1996, Pekar 2001, Shang 2004] to employing a deformable mesh whose internal energy is minimized with the distance to the closest allowed model deformation [Weese 2001, Tsaagan 2002, Kaus 2003, Heimann 2007b]. A good overview over these algorithms has recently been published by Heimann and Meinzer [Heimann 2009].

2.4.3 Implicitly Represented Shape Priors

Level sets methods describe contours or surfaces implicitly as the zero level set of a higher dimensional function. Opposite to parametric deformable models, they offer the advantage to be topologically flexible and are thus able to model highly complex anatomical structures like blood vessels or cortical surfaces. As the original level sets are not resistant to weak contour edges and suffer from a significant numerical dissipation, nowadays higher order, hybrid, and adaptive techniques are used (e.g.[Delingette 2001, Losasso 2006]) which are unfortunately less efficient and more difficult to implement than parametric models. The idea of using level sets for surface modeling was first proposed by Osher and Sethian [Osher 1988] and later used for medical image segmentation e.g. by Malladi et al. who use front propagation on stomach and artery tree structures [Malladi 1995] and Leventon et al. who additionally employ intensity and curvature priors for segmenting corpora callosa [Leventon 2000b] and by Ciofolo and Barillot who use competitive level sets for brain segmentation [Ciofolo 2005]. A thorough study about the nature of level set methods can be found in Sethian [Sethian 1999], while Osher and Paragios as well as Cremers and Deriche present elaborate overviews about applications of level set methods in the field of computer vision [Osher 2003, Cremers 2007].

In 2000, Leventon et al. proposed a segmentation algorithm where the statistics on surfaces are made directly on level-set functions [Leventon 2000a]. Since then, the idea of modeling a priori shape knowledge using level sets has gained in importance. Given a training data set of surfaces, the statistical shape prior is generated as follows: The N surface observations k in the training data set are embedded as zero level sets of the higher dimensional functions ϕ_k which are commonly represented by signed distance functions. The mean function $\bar{\phi}$ is computed by $\bar{\phi} = \frac{1}{N} \sum_{k=1}^{N} \phi_k$ and the variability model is determined by a principal component analysis done directly on the distance functions. In general, the level set segmentation is computed by a maximum a posteriori (MAP) estimation where the level set function is evolved to converge towards the boundary of the organ to be segmented. The evolution of the level set is controlled by the optimization of an energy functional which is based on the image information as well as on the statistical shape prior and additionally integrates a regularization term. This method was adapted by Tsai et al. who focused on efficiency and robustness of the algorithm [Tsai 2003] as well as by Rousson et al. who propose variational integrations of the shape prior [Rousson 2004]. In [Cremers 2006], Cremers extended the approach by dynamical priors for tracking problems.

Though, for the statistics done on the distance maps, it has to be kept in mind that the space of signed distance functions is not linear which means that a linear combination of signed distance functions does not necessarily correspond to a signed distance function. Besides, the principal components of implicit shape models describe the variability of the distance maps but not the variation of the embedded contours. Therefore, understanding the variability information on distance functions is not obvious so that it seems difficult to exploit the variability model for a physical modeling of the shape variability.

2.5   Discussion

This chapter illuminates the important role which statistical shape models play in medical imaging. Especially segmentation problems become better posed by the employment of prior shape information in the form of SSMs. Away from being a complete review on this subject, this chapter is an attempt to highlight the main approaches and to lay the ground for further research in this area. Even though SSMs have been part of the state-of-the-art for more than fifteen years, new refined SSM methods emerge every year, and several open questions remain. Especially the correspondence problem has not been solved satisfactorily in our eyes as the assumption of one-to-one correspondences on 3D surfaces seems too strong. Furthermore, most algorithms which compute SSMs employ step by step techniques by first determining correspondence, aligning the observations, computing the mean shape and finally computing the variability model. This is an intuitive technique but not a sound mathematical framework. As the mean shape and the variation modes should optimally represent the whole scene of observations, a global approach seems to be favorable where the determination of correspondence, the alignment as well as the computation of mean shape and variability are unified in one global cost function. By doing so, a theoretical convergence could be ensured. The work in this thesis will demonstrate how a statistical shape model based on correspondence probabilities can be computed in a sound mathematical scheme.

Regarding the employment of SSMs in segmentation algorithms, two independent domains were asserted: One group of methods is based exclusively on explicit representation of SSMs and segmentation contours while the other group only uses implicit SSMs and formulates implicit segmentation schemes. Naturally, both approaches feature different strengths and suffer from different weaknesses. This raises the question if and how the strict separation of the two domains could be opened in order to develop a segmentation algorithm which benefits from the advantages of both. In this thesis, it will be shown how a combination of explicit and implicit modeling could be realized which might open new insights on that matter.

CHAPTER 3

A Generative Gaussian Mixture Statistical Shape Model

Statistical shape models are a valuable tool in medical image analysis and are efficiently used in classification, recognition, reconstruction and segmentation methods. The models incorporate statistical knowledge mainly about the expected shape and shape variability. The collection of that knowledge is done by statistically evaluating the shape information of a number of observations of the respective structure. To do so, the fundamental problem of determining proper correspondence between the observations has to be solved. The solution of the correspondence problem as well as the method of model computation depends on the representation of the shapes. In this chapter, a generative method for the computation of a parametric 3D statistical shape model for point-based shape representations is developed. A probabilistic modeling is chosen instead of a deterministic one and the shapes are represented by mixtures of Gaussians. The computation of the Gaussian Mixture SSM is formulated in a generative framework.

3.1 Motivation

Most methods in the state-of-the-art compute the parameters needed for the SSM in a step-by-step manner: First, the observations are aligned in a common reference frame. Then, the mean shape is computed and finally, the variability model is determined. While usually leading to good results, the mathematical foundation is unclear and no convergence can be ensured. In order to create a sound mathematical framework, this work proposes to compute a *generative* model and unify the computation of all parameters which take part in the SSM computation into one global criterion.

Furthermore, as discussed in section 2.2, one of the central difficulties of analyzing different organ shapes in a statistical manner is the identification of correspondences between the points of the shapes. As the manual identification of landmarks is not an acceptable option in 3D, several preprocessing techniques were developed in the literature to automatically find exact one-to-one correspondences between surfaces which are represented by meshes as in [Lorenz 2000, Bookstein 1996, Styner 2003a, Vos 2004] to just name a few. A popular method is to optimize the correspondences and the registration transformation at the same time as does the Iterative Closest Points (ICP) algorithm [Besl 1992] for point clouds as explained in section 2.2.1. More elaborate methods directly combine the search of correspondences and of the SSM for a given training data set as proposed in [Zhao 2005b, Chui 2003] or the Minimum Description Length (MDL) approach to statistical shape modeling [Davies 2002c, Heimann 2005]. The MDL is used to optimize the distribution of points on the surfaces of the observations in the training data set when determining the best SSM. For unstructured point sets, the MDL approach is not suited to compute a SSM because it needs an explicit surface information. Another interesting approach proposes an entropy based criterion to find shape correspondences, but requires implicit surface representations [Cates 2006]. Other approaches combine the search for correspondences with shape based classification [Tsai 2005, Kodipaka 2007] or with shape analysis [Peter 2006b]. However, these methods are not easily adaptable to multiple observations of unstructured point sets as they either depend on underlying surface information or fix the number of points representing the surface. The approach in [Chui 2004] for unstructured point sets focuses only on the mean shape. In all cases, enforcing exact correspondences for surfaces represented by unstructured point sets leads to variability modes that not only represent the organ shape variations but also artificial variations whose importance is linked to the local sampling of the surface points.

We argue that when segmenting anatomical structures in noisy image data, the extracted surfaces (points) only represent probable surface locations. Therefore, a method for shape analysis should better rely on probabilistic point locations as presented with the rigid EM-ICP registration in [Granger 2002]. Accordingly, we propose to solve the correspondence problem by describing the observations as noisy measurements of the model. This amounts to representing the shapes by mixtures of Gaussians which are centered on the model surface points. The shapes are then aligned by maximizing the correspondence probability between all possible point pairs. It should be noted that the SoftAssign algorithm [Rangarajan 1997a] has a probabilistic formulation which is closely related but differs in that it gives the same role to the model and the observations. This is justified for a pair-wise registration but not for a group-wise model to observation registration, which is needed for the SSM computation.

This chapter is structured as follows: In section 3.2, an affine version of the EM-ICP registration algorithm is derived in order to establish a probabilistic framework for computing correspondence probabilities between the observations. Following in section 3.3, the generative Gaussian Mixture statistical shape model (GGM-SSM) is developed, and a maximum a posteriori framework is proposed to compute all model parameters and observation parameters at once. The solutions for optimizing the associated global criterion with respect to the observation and model parameters are derived in sections 3.4 and 3.5. The integration of normals as additional information into the global criterion is realized in section 3.7. We conclude this chapter with a discussion about the characteristics of the new model (section 3.8).

3.2   Expectation Maximization - ICP Algorithm

In MR or CT medical imaging, the accuracy of the anatomical representation depends on the slice thickness as well as the resolution in the plane. Even with a very high spatial resolution, partial volume effects will occur, so it has to be pointed out that the resulting image always remains an estimation of the true anatomical structure. Due to the recording techniques, there is always a certain amount of incertitude regarding the extracted image information.

For the computation of a SSM, a training data set containing segmented observations has to be created. The observations are mostly generated in a process which comprises two steps: First, an automatic, semi-automatic or manual segmentation of the respective structure is performed which results in a set of 2D binary images or one binary volume. Next, a surface extracting algorithm is applied. For both steps, a multitude of well researched and problem-adapted methods exists, nevertheless, the resulting segmentation will always be an estimation of the true structure surface. Concerning the correspondence problem, this means that the process of determining homologies between extracted surfaces relies on information which is not necessarily correct. Furthermore, one-to-one correspondences pose a problem for observations which feature distinctive shape detail differences as shown in figure 3.1. For these reasons, it is advantageous to use correspondence probabilities instead of exact correspondences. The EM-ICP algorithm is a convenient method to find those.

In this section, an affine extension for the Expectation Maximization - Iterative Closest Point registration is derived which tackles the correspondence problem by determining *correspondence probabilities* instead of one-to-one correspondences. The rigid EM-ICP was first introduced in 2002 by Granger and Pennec and proved to be robust, precise, and fast [Granger 2002]. As the aim is to model the shape variations which remain after pose, scaling and shearing variations are eliminated, an algorithm is needed which does an affine alignment of the shapes.

3.2.1   Algorithm

The EM-ICP algorithm determines the registration transformation T that best matches a model point set $M \in \mathbb{R}^{3N_m}$ onto an observation point set $S \in \mathbb{R}^{3N_s}$ with N_m and N_s describing the number of points of the model and the observation respectively. The focus lies on the *probability* of an observation point s_i to be a measure of a transformed model point $T \star m_j$. In that way, the point s_i is described as a displaced and noisy version of point m_j. Now all scene points are considered as being conditionally independent. If the point s_i corresponds exactly to the model point m_j, the measurement process can be modeled by the Gaussian probability distribution

$$p(s_i|m_j, T) \quad = \quad \frac{1}{(2\pi)^{\frac{3}{2}}|\Sigma_j|^{\frac{1}{2}}} \exp(-\frac{1}{2}(s_i - T \star m_j)^T.\Sigma_j^{-1}(s_i - T \star m_j)) \quad (3.1)$$

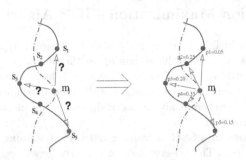

Figure 3.1: *A correspondence problem: One shape features two bumps, the other only one. How can we determine correspondences between the two? The approach used here establishes correspondence probabilities between all points representing the shape surfaces.*

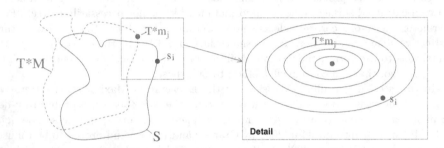

Figure 3.2: *The scene S is regarded as a set of noised measurements of the transformed model $T \star M$. The detail shows a 2D projection of the Mahalanobis distances with respect to the point $T \star m_j$. The probability of scene point s_i given T and m_j is calculated as shown in equation (3.1).*

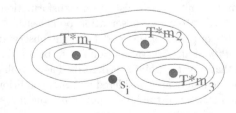

Figure 3.3: *Mixture of Gaussians describe likelihood of point s_i with respect to several model points m_j.*

where Σ_j represents the noise as the covariance of m_j. For an illustration see figure 3.2.

However, the observation point s_i can in fact be a measure of any of the model points as illustrated in figure 3.3. It is assumed that *a priori* all m_i are equally probable for being matches to s_i. Since M consists of N_m model points m_j, the probability distribution model of the spatial location of s_i is the mixture

$$p(s_i|M,T) = \frac{1}{N_m} \sum_{j=1}^{N_m} p(s_i|m_j,T). \tag{3.2}$$

Unfortunately, even under the assumption that all scene point measurements are independent, no closed form solution exists for the maximization of $p(S|M,T)$. A solution is to model the unknown correspondences $H \in \mathbb{R}^{N_s \times N_m}$ as *random hidden variables* and to maximize the log-likelihood of the complete data distribution $p(S,H|M,T)$ efficiently using the EM algorithm. We denote $E(H_{ij})$ as the expectation of point s_i being an observation of point $T \star m_j$ (with the constraint $\sum_j^{N_m} E(H_{ij}) = 1$) and compute the expectation of the log-likelihood with

$$E(\log p(S,H|M,T)) = \frac{1}{N_m} \sum_i^{N_s} \sum_j^{N_m} E(H_{ij}) \log p(s_i|m_j,T). \tag{3.3}$$

In the following, uniform priors on H are assumed.

In the **expectation step**, T is fixed and $\log p(S,H|M,T)$ is estimated to compute the expectation of correspondence $E(H)$:

$$P(H_{ij}=1) = E(H_{ij}) = \frac{\exp(-\mu(s_i,T \star m_j))}{\sum_k \exp(-\mu(s_i,T \star m_k))}$$

with $\mu(s_i,T \star m_j) = \frac{1}{2}(s_i - T \star m_j)^T.\Sigma_j^{-1}(s_i - T \star m_j)$.

In the **maximization step**, $E(H)$ is fixed and the estimated likelihood is maximized with respect to T. For this purpose, constants and normalizing factors of equation (3.3) do not have to be taken into account. Hence, the EM-ICP criterion C_{EM} to be optimized takes the following form:

$$C_{EM}(T,E) = \sum_i^{N_s} \sum_j^{N_m} E(H_{ij})(s_i - T \star m_j)^T \Sigma_j^{-1}(s_i - T \star m_j). \tag{3.4}$$

Without loss of generality, it is assumed from now on a homogeneous and isotropic Gaussian noise with variance σ^2 in order to simplify the equations. The transformation is then found by

$$\hat{T} = \operatorname*{argmin}_T \frac{1}{\sigma^2} \sum_i^{N_s} \sum_j^{N_m} E(H_{ij})\|s_i - T \star m_j\|^2. \tag{3.5}$$

We see that the elements of $E(H)$ serve as weighting factors. The solution of this least-squares estimation for a rigid transformation T can be seen in [Granger 2002].

3.2.2 Generalization to Affine Transformation

When dealing with an affine transformation T_{aff}, a point m_j is transformed by T_{aff} as follows: $T_{aff}\star m_j = Am_j + t$ with the transforming matrix $A \in \mathbb{R}^{3x3}$ and the translation vector $t \in \mathbb{R}^3$. In order to find the best translation t, equation (3.4) is differentiated with respect to t, and we obtain

$$\frac{\partial C_{EM}(t)}{\partial t} = -2\frac{1}{\sigma^2}(\sum_i^{N_s} s_i - A\sum_j^{N_m} m_j \sum_i^{N_m} E(H_{ij}) - N_s t)$$

knowing $\sum_j^{N_m} E(H_{ij}) = 1\ \forall i$. Thus, at the optimum we find

$$\hat{t} = \frac{1}{N_s}\sum_i^{N_s} s_i - A\frac{1}{N_s}\sum_j^{N_m} m_j \sum_i^{N_s} E(H_{ij}). \qquad (3.6)$$

We see that $\hat{t}$ aligns the barycentre $\bar{s} = \frac{1}{N_s}\sum_i^{N_s} s_i$ and the pseudo barycentre $\tilde{m} = \frac{1}{N_s}\sum_j^{N_m} m_j \sum_i^{N_s} E(H_{ij})$ of the two point clouds S and M. Using "barycentre" coordinates $s'_i = s_i - \bar{s}$ and $m'_j = m_j - \tilde{m}$ allows us to simplify the criterion into

$$C'_{EM}(T, E) = \frac{1}{\sigma^2}\sum_i^{N_s}\sum_j^{N_m} E(H_{ij})(s'^T_i s'_i - 2s'^T_i Am'_j + m'_j A^T Am'_j). \qquad (3.7)$$

Next, $C'_{EM}(T)$ is differentiated with respect to the affine transformation matrix A:

$$\frac{\partial C'_{EM}(A)}{\partial A} = -\frac{2}{\sigma^2}\sum_i^{N_s}\sum_j^{N_m} E(H_{ij})s'_i m'^T_j + \frac{2}{\sigma^2}\sum_i^{N_s}\sum_j^{N_m} E(H_{ij})Am'_j m'^T_j$$

$$= \frac{2}{\sigma^2}(-\Gamma + A\Upsilon)$$

with $\Upsilon, \Gamma \in \mathbb{R}^{3\times3}$.

We solve for A with

$$A\Upsilon = \Gamma \Leftrightarrow A = \Gamma\Upsilon^{-1}.$$

If Υ is singular ($det(\Upsilon) = 0$), the pseudo-inverse Υ^+ has to be determined instead of the inverse Υ^{-1}. From an implementational point of view, it is advantageous to always determine the pseudo-inverse. As Υ is symmetric, the pseudo-inverse is computed using the Jacobi method for eigenvalue decomposition. For details see section A.1.

The resulting transformation T is applied to the points of the target cloud M before the next Expectation step. The two EM-steps are alternated until $|C_{EM}(T, E)^{(i)} - C_{EM}(T, E)^{(i-1)}| < \epsilon$. A mathematical proof of convergence for the EM algorithm is provided in [Dempster 1977].

3.2.3 EM-ICP Multi-Scaling

In order to robustify the computation of the affine transformation, an iterative multi-scale scheme is implemented. Here, the variance σ^2 controlling the correspondence probabilities between shapes (as formulated in equations (3.1) and (3.2)) is used as a scale parameter. In his thesis, S. Granger analysed the influence of the variance on the convergence of the rigid EM-ICP algorithm [Granger 2003]. The results suggest that the algorithm should be started with a large variance to guarantee the robustness and that the final variance should be in the range of the real noise variance in order to ensure the most accurate results. A large variance makes sure that shape positions and rotations of source and target are aligned. A low variance makes sure that the shape details of source and target are aligned. This is implemented as follows: We start the EM-ICP registration with sigma σ_{start} in the first iteration. In each following iteration it, the sigma value is reduced to $\sigma_{it} = \text{r-factor}^{it} \cdot \sigma_{start}$ where the reduction factor is a scalar with $0 < \text{r-factor} < 1$. Its value has to be chosen carefully as a fast decrease of the multi-scale variance σ^2 could easily freeze the model in local minima. The same applies for the choice of the initial σ-value. If the sigma is chosen too small, the EM-ICP behaves like the ICP registration algorithm which means that always only one point, the closest neighbour, is fixed as corresponding point. For mathematical proof please refer to appendix A.2. If sigma is chosen too great, the source tends to shrink to the barycentre of the target. Eventually, the choice of sigma depends on the data at hand and is determined heuristically so far. In order to illustrate the influence of sigma and reduction factor in the multiscale-scheme, we examine an example: The affine EM-ICP is employed to register two kidneys represented by around 3000 points each. The value of σ_{start} is set to 12, the registration is iterated 100 times. In the first registration, no multi-scaling is performed. In the second registration, a multi-scaling is performed with a reduction factor r-factor=0.97. The algorithm with multi-scaling comes to better results as without as illustrated in figures 3.4 and 3.5.

We then test the behaviour of the affine EM-ICP on a synthetic registration problem. Our data consists of a segmented kidney S which is represented by $N = 10466$ surface points s_i and has a size of about $70mm \times 40mm \times 120mm$. We generate a second kidney S_T by deforming S with a synthetic transformation T_{synth}: $S_T = T_{synth} \star S$. Subsequently, both point sets are decimated to S^d and S_T^d using a decimation algorithm which is based on the technique presented in [Schroeder 1992]. Here, the points are splitted and moved during decimation. By choosing different decimation parameters (different number or triangles, different point priority queues) for S and S_T, it is ensured that the number of common conserved points (exact correspondences) between S^d and S_T^d is less than 15%, so real conditions - where no exact one-to-one correspondences can be determined - are simulated. Moreover, the number of points differs. In the following experiments, S^d and S_T^d are represented by around 510 points.

In order to quantify the accuracy of registration, we define a distance measure as the normalized sum of distances between all corresponding points s_i and $s_{T,i}$ of the

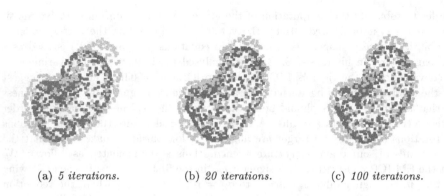

(a) *5 iterations.*   (b) *20 iterations.*   (c) *100 iterations.*

Figure 3.4: *Affine EM-ICP registration on two kidney point clouds, source in light grey and target in dark grey. The variance is set to 12 and remains constant for the whole registration process.*

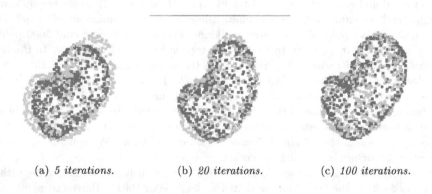

(a) *5 iterations.*   (b) *20 iterations.*   (c) *100 iterations.*

Figure 3.5: *Affine EM-ICP registration on two kidney point clouds, source in light grey and target in dark grey. The variance is set to 12 for the first iteration and is then reduced with a reduction factor of 0,97 in each new iteration.*

original, non-decimated, kidneys:

$$d^2(S, S_T) = \frac{1}{N_S} \sum_{i=1}^{N_S} \|s_i - s_{T,i}\|^2.$$

We chose this distance measure instead of comparing the computed transformation with the original one since Euclidean point distances are easier to interprete than matrix coefficient differences. In summary, the experiments are conducted by performing the following steps:

1. Choosing T_{synth} to generate S_T.

2. Decimation of S and S_T resulting in S^d and S_T^d.

3. Registration of S^d and S_T^d using the affine EM-ICP.

4. Applying the resulting transformation T_{res} to S_T.

5. Computing the distance between S and $T_{res} \star S_T$.

We tested for similarity and affine T_{synth}. The similarity transformation represents a rotation with $rot_x = 20°$, $rot_y = 10°$, and $rot_z = 5°$, a scaling of $scale_x = 1.1$, $scale_y = 0.9$, and $scale_z = 1$, and a displacement of $disp_x = 10mm$, $disp_y = 10mm$, and $disp_z = 10mm$. No shearing is applied. We start the registration with $\sigma_{start} = 8mm$ and used a reduction factor of r-factor−0.9. The algorithm converged after 30 iteration and resulted in a distance of $d(S, S_T) = 0.5mm$. The result is shown in figure 3.6.

The affine transformation has a high shearing effect with

$$T_{synth,affine} = \begin{pmatrix} 1 & 0 & 0 & 0 \\ 0.1 & 1 & 0 & 0 \\ 0.07 & 0.02 & 1 & 0 \\ 0 & 0 & 0 & 1 \end{pmatrix}.$$

Again, the registration is started with $\sigma_{start} = 8mm$ but in this experiment, the reduction factor is varied with r-factor $= \{0.5 \ 0.85 \ 0.90 \ 0.95\}$. Figure 3.8 shows the influence of the reduction factor on the convergence rate for the affine T_{synth}. The final surface distances are in the range of $d(S, S_T) = 0.35mm$ for the tested r-factors $\{0.85 \ 0.90 \ 0.95\}$. A r-factor of 0.5 however leads to a distance of $d(S, S_T) = 0.46mm$ since the algorithm freezes in a local minimum for that case. For a result example of the affine transformation experiments see figure 3.7.

We could establish that the affine EM-ICP registration results in a typical distance of $d(S, T_{res} \star S_T) \leq 0.5mm$ for our data set under the tested transformations. This value lies in the same range as the average distance of one point in S to its closest neighbour ($0.74mm$). Typically, 30 iterations sufficed for the kidney registration in this set-up. The EM-ICP needs no previous rigid registration for the affine case.

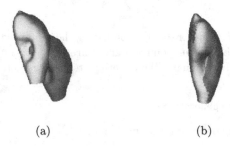

(a)   (b)

Figure 3.6: *The original objects S (dark grey) and their transformed versions S_T (light grey) (a) before registration with $d(S, S_T) = 51,7mm$ and (b) after registration with $d(S, T_{res} \star S_T) = 0.5mm$. For the EM-ICP, the kidneys were decimated from 10466 to around 510 points, we chose an initial sigma of 8mm, 30 iterations and a reducing factor of 0.9 (which leads to a final sigma of 0.38mm).*

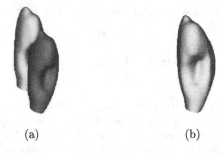

(a)   (b)

Figure 3.7: *The original objects S (dark grey) and their transformed versions S_T (light grey) (a) before registration with $d(S, S_T) = 40,3mm$ and (b) after registration with $d(S, T_{res} \star S_T) = 0.35mm$. For the EM-ICP, the kidneys were decimated from 10466 to around 510 points, we chose an initial sigma of 8mm, 30 iterations and a reducing factor of 0.9 (which leads to a final sigma of 0.38mm).*

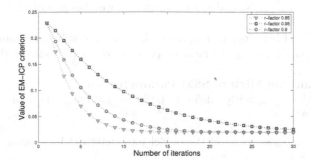

Figure 3.8: *Convergence of EM-ICP in affine kidney registration. The EM-ICP criterion values are plotted with respect to the number of iterations for three different reduction factors (r-factor). The final surface distance were all in the range of $\approx 0.35mm$. A reduction factor of 0.5 however leads to a distance of 0.46mm since the algorithm freezes in a local minimum for that case.*

3.3 The Unified Framework

In the probabilistic approach, the aim is to compute a generative model which optimally fits the given data set. We realize this by developing a global and unique criterion which is optimized iteratively with respect to all model and all observation parameters. The optimization is done through a single maximum a posteriori (MAP) criterion and leads to very efficient and closed-form solutions for (almost) all parameters *without* the need for one-to-one correspondences as is usually required by the principal component analysis. The registration transformations which are needed to match the model on the observations are computed using an affine version of the Expectation Maximization - Iterative Closest Point (EM-ICP) algorithm which is based on probabilistic correspondences and which proved to be robust and fast. By relying on correspondence probabilities, the generative statistical shape model representing the training data set is modeled as a mixture of Gaussians.

In section 3.3.1, the generative model parameters and observation parameters are presented and integrated in a unified framework. In section 3.3.2, the global criterion obtained by the MAP estimation is developed.

3.3.1 The Generative Model

We assume a training data set of segmented organs which contains N observations S_k. The observations are represented by point clouds with respectively N_k points in 3D, so that $S_k \in \mathbb{R}^{3N_k}$. We want to determine a generative statistical shape model which best represents the given observations. Here, the observations are interpreted as randomly generated by the model: The scene S_k is seen as a set of noised measurement of the model. The model itself is modeled as a random variable described by a Gaussian distribution.

In order to avoid homology assumptions, the approach is based on correspondence probabilities. In the following, the involved parameters are presented in detail.

Generative Gaussian Mixture SSM Parameters Θ:
The GGM-SSM is explicitly defined by the following 4 model parameters $\Theta = \{\bar{M}, v_p, \lambda_p, n\}$:

- $\bar{M} \in \mathbb{R}^{3N_m}$: Mean shape of the model parameterized by a point cloud of N_m points $m_j \in \mathbb{R}^3$.

- $v_p \in \mathbb{R}^{3N_m}$: n variation modes represented by N_m 3D vectors v_{pj}.

- $\lambda_p \in \mathbb{R}$: n associated standard deviations $\lambda_p \in \mathbb{R}$ which describe - similar to the classical eigenvalues of the Principal Component Analysis - the impact of the variation modes.

- n: Number of variation modes ($n \leq N$).

Observation Parameters Q:
From the parameters Θ of a given structure, the shape variations of that structure can be generated by

$$M = \bar{M} + \sum_{p=1}^{N} \omega_p v_p, \quad N \leq n$$

with $\omega_p \in \mathbb{R}$ being the deformation coefficients $\Omega = \{\omega_1, ..., \omega_n\}$ of the current shape (observation parameter) along the modes $v_1, ..., v_n$ (model parameter). The probability of obtaining a random deformed model M depends on the probability of the deformation coefficient parameters given Θ. We model the deformation coefficients distribution as Gaussian:

$$p(M|\Theta) = p(\Omega|\Theta) = \prod_{p=1}^{n} p(\omega_p|\Theta) = \frac{1}{(2\pi)^{n/2} \prod_{p=1}^{n} \lambda_p} \exp\left(-\sum_{p=1}^{n} \frac{\omega_p^2}{2\lambda_p^2}\right) \qquad (3.8)$$

where the standard deviation λ_p is a model parameter.

In the framework of the GGM-SSM computation for a training data set containing the observations S_k, the deformation coefficients are denoted ω_{kp} according to the S_k they belong to.

The second observation parameter are the registration transformations which position our system in space by aligning the model shape with each of the observations. Each transformation is associated with one observation S_k, they are denoted as $T_k = \{A_k \in \mathbb{R}^{3 \times 3}, t_k \in \mathbb{R}^3\}$ with rotational or affine matrix $A_k \in \mathbb{R}^{3 \times 3}$ and translation t_k. In order to compute the transformation which maximizes the correspondence probability between the model and a observation, the Expectation Maximization Iterative Closest Points registration algorithm which is explained in detail in section 3.2 is employed. The hidden variable in the Expectation Maximization algorithm is the correspondence probability matrix $E_{kij} \in \mathbb{R}^{N_k \times N_m}$. Its elements at position ij describe the correspondence probability for observation point s_i with model point m_j.

Applying the transformation T_k to a model point m_j is written as

$$T_k \star m_j = A_k m_j + t_k.$$

The instantiated and placed model M_k is then determined by applying the transformation to all model points m_j which is denoted as

$$M = T_k \star M. \tag{3.9}$$

We summarize the observation parameters as $Q = \{\Omega_k, T_k\}$.

The unified framework of the parameters and their specific relations are illustrated in the diagram shown in figure 3.9.

3.3.2 Optimization of Parameters through a Single MAP Criterion

As described in section 3.3.1, the approach deals with two sets of parameters:

1. **Model parameters:** $\Theta = \{\bar{M}, v_p, \lambda_p, n\}$.

2. **Observation parameters:** $Q_k = \{\Omega_k, T_k\}$ and associated nuisance parameters (hidden variables) E_k.

In order to develop a framework to compute these parameters for a given training data set S, the aim is to find the parameters Θ and Q which most probably generated that scene. The likelihood function is given by $(Q, \Theta) \mapsto p(S|Q, \Theta)$. We first approach the situation from the view point of its use, that is, it is assumed that *the model parameters in Θ are known*. We are interested in the search for the parameters linked to the shape observations: $Q = \{Q_k\}$. The model is modeled as a random variable with a Gaussian distribution which means that a prior distribution over (Q, Θ) exists which is not uniform since $p(Q, \Theta) \neq constant$. In order to take into account the prior that the model is providing on the observation parameters, a maximum a posteriori estimation should be optimized instead of a maximum likelihood estimation of Q and Θ. The posterior distribution of (Q, Θ) is then $(Q, \Theta) \mapsto p(Q, \Theta|S)$. In the MAP estimation, Bayes' theorem is used which leads to

$$\text{MAP} = -\sum_{k=1}^{N} \log(p(Q_k, \Theta|S_k)) = -\sum_{k=1}^{N} \log\left(\frac{p(S_k|Q_k, \Theta)p(Q_k|\Theta)p(\Theta)}{p(S_k)}\right). \tag{3.10}$$

Model Θ

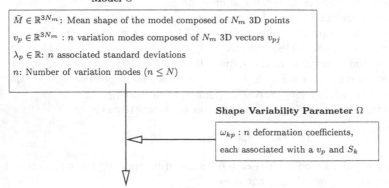

$\bar{M} \in \mathbb{R}^{3N_m}$: Mean shape of the model composed of N_m 3D points

$v_p \in \mathbb{R}^{3N_m}$: n variation modes composed of N_m 3D vectors v_{pj}

$\lambda_p \in \mathbb{R}$: n associated standard deviations

n: Number of variation modes ($n \leq N$)

Shape Variability Parameter Ω

ω_{kp} : n deformation coefficients,

each associated with a v_p and S_k

Deformation of the Model

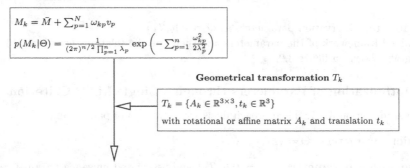

$M_k = \bar{M} + \sum_{p=1}^{N} \omega_{kp} v_p$

$p(M_k|\Theta) = \dfrac{1}{(2\pi)^{n/2} \prod_{p=1}^{n} \lambda_p} \exp\left(-\sum_{p=1}^{n} \dfrac{\omega_{kp}^2}{2\lambda_p^2}\right)$

Geometrical transformation T_k

$T_k = \{A_k \in \mathbb{R}^{3\times3}, t_k \in \mathbb{R}^3\}$

with rotational or affine matrix A_k and translation t_k

Placement in space

$M'_k = T_k \star M_k$

Correspondence probability E_k

$E_k \in \mathbb{R}^{N_k \times N_m}$

$\sum_j E_{kij} = 1$

Sampling

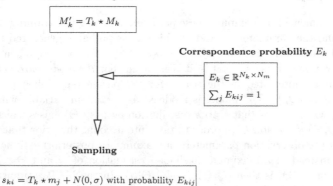

$s_{ki} = T_k \star m_j + N(0,\sigma)$ with probability E_{kij}

Figure 3.9: *Unified framework for GGM-SSM computation. The model parameters, the observation parameters and their respective relations are illustrated.*

The probability of the observations $p(S_k)$ does not depend on the model parameters Θ and $p(\Theta)$ does not play a role with Θ known. Hence, the MAP estimation can be simplified and the global criterion integrating our unified framework is the following:

$$C(Q,\Theta) = -\sum_{k=1}^{N}\left(\underbrace{\log(p(S_k|Q_k,\Theta))}_{\text{ML estimate}}+\underbrace{\log(p(Q_k|\Theta))}_{\text{Prior}}\right).$$

The first term describes a maximum likelihood (ML) estimation with $p(S_k|Q_k,\Theta) = p(S_k|T_k,\Omega_k,\Theta)$, which gives

$$p(S_k|Q_k,\Theta) = \prod_{i=1}^{N_k}\frac{1}{N_m}\sum_{j=1}^{N_m}p(s_{ki}|m_{kj},T_k) \quad \text{with} \quad m_{kj} = \bar{m}_j + \sum_{p=1}^{n}\omega_{kp}v_{pj}.$$

As a given scene point s_{ki} is modeled as a noisy measurement of a (transformed) model point m_j, the probability of the observed point is given by

$$p(s_{ki}|m_j,T_k) = \frac{1}{(2\pi)^{\frac{3}{2}}\sigma}\exp(-\frac{1}{2\sigma^2}(s_{ki}-T_k\star m_j)^T.(s_{ki}-T_k\star m_j)). \qquad (3.11)$$

The second term of $C(Q,\Theta)$ (the prior) depends on the probability of the deformation coefficients ω_{kp} as described in equation (3.8).

For the complete criterion we thus we find

$$\begin{aligned}
C(Q,\Theta) &= -\sum_{k=1}^{N}\sum_{i=1}^{N_k}\log\left(\frac{1}{N_m}\sum_{j=1}^{N_m}\frac{1}{(2\pi)^{\frac{3}{2}}\sigma}\exp\left(-\frac{\|s_{ki}-T_k\star m_{kj}\|^2}{2\sigma^2}\right)\right) \\
&\quad +\sum_{k=1}^{N}\left(\log((2\pi)^{n/2})+\log(\sum_{p=1}^{n}\lambda_p)+\sum_{p=1}^{n}\frac{\omega_{kp}^2}{2\lambda_p^2}\right) \qquad (3.12) \\
&= \alpha(n)+\beta(N_m)-\zeta(\sigma)+\sum_{k=1}^{N}C_k(Q_k,\Theta).
\end{aligned}$$

The number of variation modes is not optimized but a fixed number is assumed. The number N_m of points in the model is fixed and a multi-variance scheme is employed. Hence, $\alpha(n) = \sum_k\log((2\pi)^{n/2})$, $\beta(N_m) = \sum_k N_k\log(N_m)$ and $\zeta(\sigma) = NN_k\log\left((2\pi)^{-\frac{3}{2}}\sigma^{-1}\right)$ become constants.

Our criterion thus simplifies to $C_{global}(Q,\Theta) = \sum_{k=1}^{N}C_k(Q_k,\Theta)$ with

$$C_k(Q_k,\Theta) = \sum_{p=1}^{n}\left(\log(\lambda_p)+\frac{\omega_{kp}^2}{2\lambda_p^2}\right)-\sum_{i=1}^{N_k}\log\left(\sum_{j=1}^{N_m}\exp\left(-\frac{\|s_{ki}-T_k\star m_{kj}\|^2}{2\sigma^2}\right)\right). \qquad (3.13)$$

The first term is responsible for maximizing the probability of deformation while the second term tries to minimize the point distances of model and observations. The global criterion of equation (3.13) incorporates the unified framework for the model computation. By optimizing it alternately with respect to the operands in $\{Q, \Theta\}$, we are able to determine all parameters we are interested in.

Some terms will recur in the different optimizations as the derivative of the second term of the global criterion is always performed in the same manner. We will introduce the following notations for simplification reasons: The derivative of an arbitrary function ξ

$$\xi_{kij}(T_k, \Omega_k, \bar{M}, v_p, \lambda_p) = \log \sum_{j=1}^{N_m} \exp \left(-\frac{\|s_{ki} - T_k \star m_{kj}\|^2}{2\sigma^2} \right)$$

with respect to one of the function's parameters (let's say x) is

$$\frac{\partial \xi_{kij}}{\partial x} = -\sum_{j=1}^{N_m} \gamma_{kij} \frac{(s_{ki} - T_k \star m_{kj})^T}{\sigma^2} \ \frac{\partial(s_{ki} - T_k \star m_{kj})}{\partial x}$$

with

$$\gamma_{kij} = \frac{\exp \left(-\frac{\|s_{ki} - T_k \star m_{kj}\|^2}{2\sigma^2} \right)}{\sum_{l=1}^{N_m} \exp \left(-\frac{\|s_{ki} - T_k \star m_{kl}\|^2}{2\sigma^2} \right)}. \tag{3.14}$$

The details of this derivative can be found in appendix A.3.

Note that the variable γ_{kij} is equal to the elements E_{kij} of the expectation matrix which means that the derivatives of all parameters are weighted by the correspondence probabilities of all s_{ki} and m_j.

3.4 Computation of the Observation Parameters

In this section, the alternated optimizations of the observation parameters $\{T_k, \Omega_k\}$ with fixed and known model parameters $\Theta = \{\bar{M}, v_p, \lambda_p, n\}$ are described in detail.

3.4.1 Transformation

We optimize the global criterion (equation (3.13)) with respect to the spatial transformation T_k, so Ω_k and Θ are fixed.

Here, the concept of the affine EM-ICP registration described elaborately in section 3.2 is used where the correspondence probabilities E_{kij} are modeled as hidden variables.

1. The Expectation Step:
In the expectation step, the transformation T_k is fixed. We compute the expectancy of the log-likelihood of the complete data distribution and derive

$$E_{kij} = \gamma_{kij} = \frac{\exp\left(-\frac{\|s_{ki} - T_k \star m_{kj}\|^2}{2\sigma^2}\right)}{\sum_{l=1}^{N_m} \exp\left(-\frac{\|s_{ki} - T_k \star m_{kl}\|^2}{2\sigma^2}\right)}, \tag{3.15}$$

compare equation (3.14).

2. The Maximization Step:
In the maximization step, the correspondence probabilities E_k are fixed, and the transformations T_k have to be determined. Therefore, the global criterion is optimized first with respect to the translation t_k and next with respect to the affine registration matrix A_k.

Optimization with respect to the translation
We optimize the criterion with respect to the translation t_k. For the derivative of the second term, the general derivative described in equation (3.14) is employed:

$$\frac{\partial C_k(Q_k, \Theta)}{\partial t_k} = +\sum_{i=1}^{N_k}\sum_{j=1}^{N_m} \gamma_{kij} \frac{(s_{ki} - T_k \star m_{kj})^T}{\sigma^2} \frac{\partial(s_{ki} - T_k \star m_{kj})}{\partial t_k}$$

with

$$\frac{\partial(s_{ki} - T_k \star m_{kj})}{\partial t_k} = \frac{\partial}{\partial t_k}(s_{ki} - t_k - A_k(\bar{m}_j + \sum_{p=1}^{n} \omega_{kp} v_{pj})) = -I_{3 \times 3}.$$

Solving for $\frac{\partial C_k(Q_k, \Theta)}{\partial t_k} = 0$, we find

$$\frac{1}{\sigma^2}\sum_{i=1}^{N_k}\sum_{j=1}^{N_m} \gamma_{kij}(s_{ki} - t_k - A_k(\bar{m}_j + \sum_{p=1}^{n} \omega_{kp} v_{pj})) = 0$$

which gives explicitly the transformation

$$t_k = \tilde{s}_k - A_k\left(\tilde{\bar{m}}_j + \sum_{p=1}^{n} \omega_{kp} \tilde{v}_p\right). \tag{3.16}$$

with

$$\tilde{s}_k = \frac{1}{N_k}\sum_{i=1}^{N_k} s_{ki}, \quad \tilde{\bar{m}}_j = \frac{1}{N_m}\sum_{j=1}^{N_m} \bar{m}_j \sum_{i=1}^{N_k} \gamma_{kij} \quad \text{and} \quad \tilde{v}_p = \frac{1}{N_k}\sum_{i=1}^{N_k} \gamma_{kij} v_{pj}.$$

This is no more than the superposition of weighted barycentres with weights a bit more complex than usual since the model barycentre includes a correction for the modes.

Optimization with respect to the affine matrix
In order to optimize the criterion with respect to the affine matrix A_k, the translation t_k is replaced as found above (equation (3.16)), so the implementation of the whole transformation derivative becomes simpler. The points of the shapes are now expressed with respect to their barycentres and we set

$$s'_{ki} = s_{ki} - \tilde{s}_k \quad \text{and} \quad m'_{kj} = \bar{m}_j - \tilde{\bar{m}}_j + \sum_{p=1}^{n} \omega_{kp}(v_{pj} - \tilde{v}_p).$$

The first term of the global criterion in equation (3.13) does not contain transformation parameters, so we can rewrite our criterion to

$$C'_k(Q_k, \Theta) = \text{const} - \sum_{k=1}^{N} \sum_{i=1}^{N_k} \log \left(\sum_{j=1}^{N_m} \exp \left(-\frac{\|s'_{ki} - A_k m'_{kj}\|^2}{2\sigma^2} \right) \right).$$

Then the derivative of $C'_k(Q_k, \Theta)$ is solved with respect to A_k. Here, the derivative form shown in equation (A.2) is used which simply is:

$$\frac{\partial C'_k(Q_k, \Theta)}{\partial A_k} = -\sum_{i=1}^{N_k} \sum_{j=1}^{N_m} \gamma_{kij} \frac{\partial}{\partial A_k} \frac{\|s'_{ki} - A_k m'_{kj}\|^2}{2\sigma^2} = 0$$

and which finally leads to a matrix equation in the form of

$$A_k \sum_{i=1}^{N_k} \sum_{j=1}^{N_m} \gamma_{kij} m'_{kj} m'^T_{kj} = \sum_{i=1}^{N_k} \sum_{j=1}^{N_m} \gamma_{kij} s'_{ki} m'^T_{kj}$$

$$\Leftrightarrow A_k \Upsilon_k = \Psi_k, \quad \Upsilon_k, \Psi_k \in \mathbb{R}^{3 \times 3}.$$

(The detailed derivation can be found in appendix A.3.) The elements of Υ_k and Ψ_k in row r and column s are determined by

$$\upsilon[r][s] = \sum_{i=1}^{N_k} \sum_{j=1}^{N_m} \gamma_{kij} \, m'_{kj}[r] \, m'_{kj}[s]$$

and

$$\psi[r][s] = \sum_{i=1}^{N_k} \sum_{j=1}^{N_m} \gamma_{kij} \, s'_{ki}[r] \, m'_{kj}[s].$$

where $m'_{kj}[s]$ denotes the entry of vector m'_{kj} at position s.

Hence, the computation of the transformation can be efficiently done in a closed-form solution by solving a 3×3 equation system.

3.4.2 Deformation Coefficients

In order to compute the deformation coefficients $\Omega = \{\Omega_k\}$, the global criterion (equation (3.13)) is optimized with respect to the deformation coefficients Ω_k. The transformations T_k and the model parameters Θ are fixed. For the derivative of the second term of the criterion, again the general derivative described in equation (3.14) is employed. For details please see appendix A.3. We finally find

$$\frac{\partial C_k(Q_k, \Theta)}{\partial \omega_{kp}} = \frac{\omega_{kp}}{\lambda_p^2} - \frac{1}{\sigma^2} \sum_{i=1}^{N_k} \sum_{j=1}^{N_m} \gamma_{kij}(s_{ki} - T \star m_{kj})^T A_k v_{pj} = 0.$$

In order to simplify, let us introduce the real values d_{kp} and g_{kqp} (with $g_{kqp} = g_{kpq}$):

$$d_{kp} = \sum_{i=1}^{N_k} \sum_{j=1}^{N_m} \gamma_{kij}(s_{ki} - t_k - A_k \bar{m}_j)^T A_k v_{pj}$$

and

$$g_{kqp} = \sum_{i=1}^{N_k} \sum_{j=1}^{N_m} \gamma_{kij} v_{qj}^T A_k^T A_k v_{pj}.$$

Thus, the system which has to be solved for the optimal ω_{kp} is (for all p):

$$\frac{\sigma^2}{\lambda_p^2} \omega_{kp} - d_{kp} + \sum_{q=1}^{n} \omega_{kq} g_{kqp} = 0.$$

We solve this equation with respect to all ω_{kp} at a time by switching to a matrix notation where all ω_{kp} are sorted in vector $\Omega_k \in \mathbb{R}^n$, all d_{kp} are sorted in vector $\vec{d}_k \in \mathbb{R}^n$ and all g_{kpq} are sorted in the symmetric matrix $G_k \in \mathbb{R}^{n \times n}$:

$$0 = \sigma^2 \begin{pmatrix} \frac{1}{\lambda_1^2} & & 0 \\ & \ddots & \\ 0 & & \frac{1}{\lambda_n^2} \end{pmatrix} \Omega_k - \vec{d}_k + G_k \Omega_k.$$

$$\Leftrightarrow \quad (G_k + R_{nn}) \Omega_k = \vec{d}_k \tag{3.17}$$

with matrix $R_{nn} = \sigma^2 diag(\lambda_1^{-2}, ..., \lambda_n^{-2})$. In order to compute the ω_{kp}, for each k the matrix G_k and the vector $\vec{d}_k$ have to be evaluated. In the implementation, the linear equation system is solved using a LU decomposition of $(G_k + R_{nn})$.

3.5 Computation of the Model Parameters

For the computation of all model parameters, we assume the observation parameters $Q_k = \{\Omega_k, T_k\}$ to be fixed and known and optimize the global criterion of equation (3.13) with respect to the parameters in Θ with $\Theta = \{\bar{M}, v_p, \lambda_p\}$.

3.5.1   Mean Shape

We optimize the global criterion (equation (3.13)) with respect to the mean shape $\bar{M}$, so the standard deviation λ_p, the variation modes v_p and the observation parameters Q_k are fixed. We evaluate the derivative for each mean shape point $\bar{m}_j$. The first term of the global criterion in equation (3.13) does not contain any m_j, so we concentrate on the second term. Using the general derivative presented in equation (3.14), we directly find

$$\frac{\partial C_{global}(Q,\Theta)}{\bar{m}_j} = + \sum_{k=1}^{N} \sum_{i=1}^{N_k} \gamma_{kij} \frac{(s_{ki} - T_k \star m_{kj})^T}{\sigma^2} \; \frac{\partial(s_{ki} - T_k \star m_{kj})}{\partial \bar{m}_j} = 0.$$

We finally solve for m_j by

$$\bar{m}_j = \left(\sum_{k=1}^{N} \sum_{i=1}^{N_k} \gamma_{kij} A_k^T A_k \right)^{-1} \sum_{k=1}^{N} \sum_{i=1}^{N_k} \gamma_{kij} A_k^T (s_{ki} - t_k - A_k \sum_{p=1}^{n} \omega_{kp} v_{pj}) \qquad (3.18)$$

which is derived in detail in appendix A.3. We see that the mean shape points are computed as the average of all observation points which are weighted by their respective correspondence probabilities γ_{kij}.

3.5.2   Standard Deviation

We optimize the global criterion (equation (3.13)) with respect to the standard deviation λ_p, so $\bar{M}, v_p$ and Q_k are fixed. The derivative in this case is quite easy:

$$\frac{\partial C_{global}(Q,\Theta)}{\partial \lambda_p} = \sum_{k=1}^{N} \left(\frac{1}{\lambda_p} - \frac{\omega_{kp}^2}{\lambda_p^3} \right) = 0$$

$$\Leftrightarrow \quad \lambda_p^2 = \frac{1}{N} \sum_{k=1}^{N} \omega_{kp}^2. \qquad (3.19)$$

This is consistent with the ML estimation of the standard deviation based on a normal distribution.

3.5.3   Variation Modes

We optimize the global criterion (equation (3.13)) with respect to the variation modes v_p, so all $\lambda_p, \bar{M}$ and Q_k are fixed. As we are working with a matrix notation, we first define the matrix $V \in \mathbb{R}^{3N_m \times n}$ containing the variation modes $v_p \in \mathbb{R}^{3N_m}$ in its columns. The computation of the variation modes is complex, for one as is is has to be made sure that the resulting vectors are orthogonal to each other:

$$v_p^T v_q = \delta_{pq} = \begin{cases} 1 & \text{if } p = q \\ 0 & \text{if } p \neq q \end{cases}$$

which leads to the constraint

$$V^T V = I_{n \times n}.$$

In order to integrate this constraint into the optimization, we employ Lagrange multipliers. This means that a new variable $Z \in \mathbb{R}^{n \times n}$ is introduced with a Lagrange function Λ where

$$\frac{\partial \Lambda}{\partial Z} = 0 \quad \Leftrightarrow \quad V^T V = I_{n \times n}$$

and our global criterion is extended to

$$\Lambda = C_{global} + \frac{1}{2} tr\left(Z(V^T V - I_{n \times n})\right). \tag{3.20}$$

We differentiate the two terms independently and point-wise. Here, $v_{jp} \in \mathbb{R}^3$ denote the elements of v_p which model the variation of model point m_j. We begin with the derivative of C_{global}. :

$$\frac{\partial C_{global}}{\partial \vec{v}_{jp}} = -\frac{1}{\sigma^2} \sum_{k=1}^{N} \sum_{i=1}^{N_k} \gamma_{kij}(s_{ki} - T_k \star m_{kj})^T \, \omega_{kp} A_k$$

In order to simplify the notation for clarity purposes, in the following we denote

$$\frac{\partial C_{global}}{\partial \vec{v}_{jp}} = \sum_{q=1}^{n} B_{pqj} \vec{v}_{jq} - \vec{q}_{jp}$$

with

$$\vec{q}_{jp} = \frac{1}{\sigma^2} \sum_{k=1}^{N} \sum_{i=1}^{N_k} \gamma_{kij}(s_{ki} - t_k - A_k \bar{m}_j)^T \, \omega_{kp} A_k, \quad q_{jp} \in \mathbb{R}^3$$

and

$$B_{pqj} = \frac{1}{\sigma^2} \sum_{k=1}^{N} \sum_{i=1}^{N_k} \gamma_{kij} \omega_{kq} \omega_{kp} A_k^T A_k, \quad B_{pqj} \in \mathbb{R}^{3 \times 3} \quad \forall j.$$

Differentiating the Lagrange multiplier with respect to $\vec{v}_{jp}$ gives

$$\frac{\partial}{\partial \vec{v}_{jp}} \frac{1}{2} tr\left(Z(V^T V - I_{n \times n})\right) = \frac{\partial}{\partial \vec{v}_{jp}} \frac{1}{2} tr\left(Z V^T V\right)$$

$$= \sum_{q=1}^{n} \frac{1}{2}(z_{qp} + z_{pq}) \vec{v}_{jq} \text{ with } z_{qp} = z_{pq}.$$

We now summarize the derivative to

$$\frac{\partial \Lambda}{\partial \vec{v}_{jp}} = \sum_{q=1}^{n} z_{qp} \vec{v}_{jq} + \sum_{q=1}^{n} B_{pqj} \vec{v}_{jq} - \vec{q}_{jp}. \tag{3.21}$$

In the *rigid case*, A_k is a rotation matrix - and thus orthonormal - so it holds $A_k^T A_k = I_{3\times3}$. The matrix B_{pqj} can then be written as the identity matrix multiplied by a scalar: $B_{pqj} = b_{pqj} I_{3\times3}$ with $b_{pqj} = \frac{1}{\sigma^2} \sum_{k=1}^{N} \sum_{i=1}^{N_k} \gamma_{kij} \omega_{kq} \omega_{kp}$. Hence we can simplify the solution of $\frac{\partial \Lambda}{\partial \vec{v}_{jp}} = 0$ to a vector summation:

$$\sum_{q=1}^{n} \left(z_{qp} I_{3\times3} + b_{pqj} I_{3\times3} \right) \vec{v}_{jq} = \vec{q}_{jp} \quad \Leftrightarrow \quad \sum_{q=1}^{n} \vec{v}_{jq} (z_{qp} + b_{pqj}) = \vec{q}_{jp} \tag{3.22}$$

This equation cannot be extended to a matrix notation in order to compute all $\vec{v}_{jp}$ at the same time because we deal with a different b_{pqj} for each point j, thus, B would be a tensor. Therefore, we approach the problem regarding each band $[V]_{\{j\}} \in \mathbb{R}^{3\times n}$ of matrix $V \in \mathbb{R}^{3N_m \times n}$ separately with

$$[V]_{\{j\}} = [\vec{v}_{j1}, ..., \vec{v}_{jq}, ..., \vec{v}_{jn}].$$

There are N_m bands $[V]_{\{j\}}$.

Now we can write equation (3.22) in a matrix notation

$$[V]_{\{j\}} (B_j + Z) = [Q]_{\{j\}}.$$

with the matrix $B_j \in \mathbb{R}^{n\times n}$ holding the b_{pqj} and the matrix $[Q]_{\{j\}} \in \mathbb{R}^{3\times n}$ holding the $\vec{q}_{jp}$. The computation of each band $[V]_{\{j\}}$ is then realized in an iterative manner as follows:

1.) If Z is known we can compute V: $[V]_{\{j\}} = [Q]_{\{j\}} (B_j + Z)^{-1}$.

2.) If all $[V]_{\{j\}}$ are known, we can determine Z: $[V]_{\{j\}} Z = [Q]_{\{j\}} - [V]_{\{j\}} B_j \ \forall j$.

For readability reasons, we set $[Q]_{\{j\}} - [V]_{\{j\}} B_j = [\tilde{Q}]_{\{j\}}$. Looking at all j simultaneously, we find the following matrix equation

$$VZ = \tilde{Q}.$$

with $V \in \mathbb{R}^{3N_m \times n}$, $Z \in \mathbb{R}^{n\times n}$ and $\tilde{Q} \in \mathbb{R}^{3N_m \times n}$.

For the implementation, we add two steps. First, we force the V resulting from step 1.) to be orthonormal. To do so, we apply first a singular value decomposition $V = USR^T$ with $U \in \mathbb{R}^{3N_m \times n}$, $S \in \mathbb{R}^{n\times n}$ and $R \in \mathbb{R}^{n\times n}$. Then we replace V with its orthonormal parts $V \leftarrow UR^T$.

Next, we want Z to be symmetric. Hence, instead of computing $Z = V^T \tilde{Q}$ we compute

$$Z = \frac{1}{2} \left(V^T \tilde{Q} + (V^T \tilde{Q})^T \right).$$

Finally, the optimization of the global criterion with respect to $\vec{v}_{jp}$ is done as follows:

We iterate

1. Compute $\tilde{Q}$ with bands $[\tilde{Q}]_{\{j\}} = [Q]_{\{j\}} - [V]_{\{j\}} B_j$.

2. Compute $\tilde{Z} = V^T \tilde{Q}$ and $Z = \frac{1}{2}(\tilde{Z} + \tilde{Z}^T)$.

3. Update V band per band: $[V]_{\{j\}} = [Q]_{\{j\}} (B_j + Z)^{-1}$.

4. Modify $V = USR^T$ to be orthonormal: $V \leftarrow UR^T$.

until $\|V^{t+1} - V^t\|^2 \leq \epsilon$.

In the *affine case*, it holds $A_k^T A_k \neq I_{3\times 3}$, so the solution to $\frac{\partial \Lambda}{\partial \vec{v}_{jp}} = 0$ is a bit more cumbersome as B_{pqj} is not a diagonal matrix anymore and not sparse. In the following, the general approach is explained. For all j and all p we want to solve

$$\sum_{q=1}^{n} \left(z_{qp} I_{3\times 3} + B_{pqj}\right) \vec{v}_{jq} = \vec{q}_{jp} \quad \Leftrightarrow \quad \sum_{q=1}^{n} \tilde{B}_{pqj} \vec{v}_{jq} = \vec{q}_{jp}. \tag{3.23}$$

For a matrix notation, we arrange the elements of the variation modes v_p in the vectors $[\hat{V}]_{\{j\}} \in \mathbb{R}^{3n}$ with

$$[\hat{V}]_{\{j\}} = \begin{pmatrix} \vec{v}_{j1} \\ \vdots \\ \vec{v}_{jq} \\ \vdots \\ \vec{v}_{jn} \end{pmatrix}.$$

Then we arrange the matrices $\hat{B}_{pqj}$ in $[B_j]_{pq} \in \mathbb{R}^{3n \times 3n}$:

$$[B_j]_{pq} = \begin{pmatrix} \hat{B}_{11j} & \cdots & \hat{B}_{1qj} & \cdots & \hat{B}_{1nj} \\ \vdots & \ddots & \vdots & \ddots & \vdots \\ \hat{B}_{p1j} & \cdots & \hat{B}_{pqj} & \cdots & \hat{B}_{pnj} \\ \vdots & \ddots & \vdots & \ddots & \vdots \\ \hat{B}_{n1j} & \cdots & \hat{B}_{nqj} & \cdots & \hat{B}_{nnj} \end{pmatrix},$$

so we obtain the following linear system to solve:

$$[B_j]_{pq}[\hat{V}]_{\{j\}} = [\hat{Q}]_{\{j\}}$$

Again we realize the computation iteratively by solving alternately for Z and for V. In practice, after a first rough alignment of the observations, the values of $A_k^T A_k$ come close to the identity matrix, so the rigid variant of the variation mode computation can be employed which is faster.

3.6 Practical Aspects

3.6.1 Initialization and Control of the Parameters

As the computation of the observation parameters is based on known model parameters $\Theta = \{\bar{M}, v_p, \lambda_p\}$, the mean shape $\bar{M}$ is initialized with one of the observations S_k in the given data set, preferably with a typical shape. Next, by applying the EM-ICP registration, the resulting correspondence probabilities between $\bar{M}$ and each S_k are evaluated, and "virtual" one-to-one correspondences are determined. We introduce the virtual corresponding points $\check{s}_{kj}$ for each m_j and each S_k by evaluating the mean position of the probabilistic correspondences:

$$\check{s}_{kj} = \sum_i^{N_s} \frac{E(H_{k_{ij}})}{\sum_i E(H_{k_{ij}})} (T_k^{-1} \star s_{ik}). \tag{3.24}$$

The $\check{s}_{kj}$ represent probable sampling points of an unknown underlying surface of observation S_k. We compute a set of $\check{s}_{kj}$ for each S_k. The resulting sets of assumed exact correspondences $(T \star m_j, \check{s}_{kj})$ are then used as input for the Principal Components Analysis to compute the initial eigenvectors v_p and the initial eigenvalues λ_p. For a detailed explanation of the computation see section 3.6.2. The observation parameters $Q = \{T, \Omega\}$ are initialized with $A_k = I_{3\times3}$ and $t_k = (0, 0, 0)$ for all k for the transformation and with $\omega_{kp} = 0$ for all k and all p.

In order to test for the sensibility of our SSM computation with respect to the initial mean shape, we compared the mean shape results which are obtained when using dissimilar initial mean shapes M_1 and M_2. We established that M_1 can be generated based on the SSM found with M_2 with statistically very small deformation coefficients ω_{1p}: $M_1 = M_2 + \sum_p \omega_{1p} \vec{v}_p$ with $\omega_{1p} << \lambda_{2p}$ [Hufnagel 2007b].

As the aim is to find a good balance between complexity and simplicity of the model, the dimension of the variation mode vector space is reduced during the iterated computation of the parameters. If the standard deviation λ_p becomes "too small", the associated variation modes v_p are no longer taken into account. This does scarcely influence the convergence rate of the global criterion as shown in figure 3.10.

3.6.2 Solving for the Initial Variation Modes

A training data set containing N observations S_k with a fixed number N_m of virtual corresponding points is cleared of the mean and then stored in the matrix $B \in \mathbb{R}^{3N_m \times N}$. In order to compute the principal components, the associated covariance matrix is built with $Cov(B) = BB^T \in \mathbb{R}^{3N_m \times 3N_m}$, and a eigenvalue decomposition is performed:

$$BB^T = ESE^T$$

where $S \in \mathbb{R}^{3N_m \times 3N_m}$ is a diagonal matrix which contains the eigenvalues of BB^T and $E \in \mathbb{R}^{3N_m \times 3N_m}$ is an orthogonal matrix containing the associated eigenvectors.

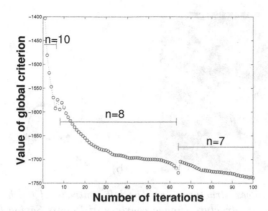

Figure 3.10: *Global criterion values of SSM computation for synthetic ellipsoid data set as illustrated in section 4.2.1.1. Since variation modes whose standard deviation fall below a certain threshold are discarded, the number n of variation modes diminishes from 10 to 7 during computation.*

However, for representing an organ like e.g. the kidney with a reasonable amount of details, at least $N_m = 3000$ points (if evenly distributed) are necessary, thus, the system to solve becomes very large with $Cov(B) \in \mathbb{R}^{9000 \times 9000}$ and is not sparse. Therefore, we apply an alternative solution to the standard eigenvalue decomposition and employ the Singular Value Decomposition (SVD) of B:

$$B = U\Sigma V^T \tag{3.25}$$

with U being an orthogonal matrix $U \in \mathbb{R}^{3N_m \times 3N_m}$, V^T being the transpose of the orthogonal matrix $V \in \mathbb{R}^{N \times N}$ and Σ being a diagonal matrix $\Sigma \in \mathbb{R}^{mxn}$ with the singular values σ_i on the diagonal. Now we use these components to represent BB^T resulting in

$$BB^T = U\Sigma V^T V \Sigma^T U^T = U\Sigma\Sigma^T U^T = ESE^T. \tag{3.26}$$

We see that U holds the sought eigenvectors of the big system as $U = E$ while $\Sigma\Sigma^T$ hold the eigenvalues of the covariance matrix. Using the singular value decomposition means that we never need the space $3N_m \times 3N_m$ to compute the covariance matrix. Moreover, the SVD is numerically more stable than the eigenvalue decomposition and therefore more accurate if the covariance matrix is ill-conditioned [Kalman 1996]. For a detailed derivation of eigenvalue and singular value decomposition please refer to section A.1.

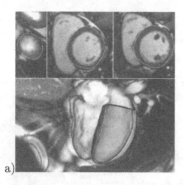
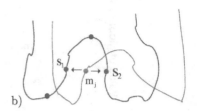

Figure 3.11: *Non-convex structures. a) The left ventricle of the heart is an example for a non-convex organ structure (image courtesy of Dennis Säring [Säring 2009]). b) Synthetic examples: Points which lie close to one another do not necessarily belong to the same part of one shape. More information than the Mahalanobis distance is needed in order to determine the correct correspondence for point m_j in this illustrated case.*

3.7 Extension of the Criterion for Non-Convex Structures

The EM-ICP algorithm works very well for shapes which are convex. Concave shapes however pose a problem as points which lie close to one another do not necessarily belong to the same part of the shape. However, their correspondence probability will be high according to the EM-ICP. For an example see figure 3.11 which shows the left ventricle of the heart and an illustrative synthetic structure.

3.7.1 Integration of Normals

For non-convex shapes, an additional information is needed about the shape alongside the Mahalanobis distances used in the EM-ICP. When looking at the figure 3.11, what easily comes to mind is the distinction of the direction the surface is facing. Therefore, the normal information is integrated into the global criterion to obtain small probabilities of correspondence between points which feature normals showing in very different directions.

Let us denote the normalized normal belonging to point s_i as η_{si} and the normalized normal belonging to point m_j as η_{mj}. We could now either measure the difference between the normals by analysing the angle between them or just by using the Euclidean norm $\|\eta_{si} - \eta_{mj}\|$. Before comparing the normals, the transformation T has to be applied to the normal vector. This is done by multiplying the inverted and transposed transformation matrix with the normal vector. The translation is not needed: $T \star \eta_{mj} = (A^{-1})^T \eta_{mj}$. Next, a renormalization of the normal is done, so in our case $T \star \eta_{mj} = \frac{(A^{-1})^T \eta_{mj}}{|(A^{-1})^T \eta_{mj}|}$. A small difference means high probability, so we extend the term of the EM-ICP given in equation (3.11) to obtain the correspondence probability of point s_i with respect to the transformed point m_j by

$$p(s_i|T, m_j) = \frac{1}{const} \exp(-\frac{\|s_i - T \star m_j\|^2}{2\sigma^2}) \exp(-\frac{\|\eta_{si} - T \star \eta_{mj}\|}{2\sigma_\eta^2}).$$

The correspondence probability relying on additional (normal) information between two points can be directly integrated in the global criterion. The elements of the expectation matrix and therefore the values γ_{kij} in the derivatives simply change to

$$\gamma_{kij}^\eta = \frac{\exp(-\frac{\|s_{ki} - T_k \star m_{kj}\|^2}{2\sigma^2} - \frac{\|\eta_{ski} - T_k \star \eta_{mkj}\|}{2\sigma_\eta^2})}{\sum_{l=1}^{N_m} \exp\left(-\frac{\|s_{ki} - T_k \star m_{kl}\|^2}{2\sigma^2} - \frac{\|\eta_{ski} - T_k \star \eta_{mkj}\|}{2\sigma_\eta^2}\right)}.$$

Only the computation of the transformation matrix becomes more complicated as the derivative of the normal term has to be taken into account.

3.7.2 Estimating Normals for Unstructured Point Clouds

The computation of normal vectors for a continuous surface is straightforward. However, the computation of normals for a non-oriented unstructured point cloud proves to be more difficult as no connectivities between the points exist. Therefore, additional information as connectivity or tangential planes have to be estimated.

Often, numerical techniques as first proposed in [Hoppe 1992] and then extended in e.g. [Pauly 2003, Mitra 2004] are used. Basically, for each point in the point cloud a normal is estimated by first computing a tangential plane which is obtained by applying the Least-Squares method to the k nearest neighbours. The normal is then computed as the vector perpendicular to that plane. Another main approach is a combinatorial one based on Voronoi/Delaunay properties as proposed by [Amenta 1999] for noise-free data and then extended by e.g. [Dey 2004] to noisy data.

An interesting approach computes the normals in a probabilistic framework as shown in [Granger 2003]. It is based on the aspect that the space of normals forms a differential manifold analogous to a sphere. The computation of normals for an unstructured point cloud is then done following a rigorous mathematical notion on random normal statistics [Pennec 1996]. The probability for a normal $\vec{n}_s$ at point s knowing the position of a neighbouring point s_i at distance d is given by $p(\vec{n}_s|s, s_i) = p(|\phi|, d)$ with ϕ being the angle between the normal and the segment ss_i. For an illustration see figure 3.12. This probability is synthesized by a tensor formulation and finally leads to the following algorithm for computing all normals of a point cloud:
For each point s_i:

- Determine a number of closest neighbours s_j using a kD-tree.

- Compute the tensor $T = \sum_j \exp(-4a^2|s_is_j|)\frac{s_is_j}{|s_is_j|}(\frac{s_is_j}{|s_is_j|})^T$ where a^2 represents the angular dispersion of the normal for a distance of $1mm$.

Figure 3.12: *The most probable normal direction for point s is computed knowing the positions of the neighbours s_i.*

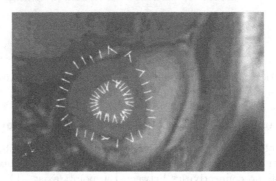

Figure 3.13: *Estimation of normals using image information.*

- Determine eigenvectors and eigenvalues of T.

- Normal $\vec{n}_{s_i}$ equals eigenvector with greatest eigenvalue.

Another feasible approach for establishing normal information is to exploit image information of the observations if available. For organs whose grey values at the boundary clearly differ from those of the background, a gradient image is computed. Following that, a normal is automatically estimated for each point of the observation based on the gradient information. An example is illustrated for the approximation of normals for the left ventricle in an MR image, see figure 3.13.

3.8   Discussion

In this chapter, a novel algorithm was developed to compute a generative Gaussian Mixture statistical shape model which is based on a sound mathematical framework. The computation of the SSM is realized as an optimization problem: An algorithm is proposed to optimize for model parameters and observation parameters through a single maximum a posteriori criterion which led to a mathematically sound and unified framework. Closed form solutions were effectually derived for optimizing the associated criterion alternately for almost all parameters. From a theoretical point of view, a very powerful feature of the method is that we are optimizing a unique criterion. Thus, theoretically the convergence is ensured. In practice, the convergence rate has to be adapted to the problem at hand as e.g. a too fast decrease of the multi-scale variance σ^2 might freeze the model in local minima. As opposed to most approaches in the literature, no principal component analysis is employed. SSM computation methods which rely on one-to-one correspondences and perform a PCA on the associated covariance matrix compute a number of eigenmodes which model both shape variation and noise. In order to discard the noise-related variations from the final variability model, eigenmodes with small eigenvalues are not taken into account. This is largely an heuristic method. In contrast, in the presented GGM-SSM the variation modes only model the shape variation as the noise is represented separately through the Gaussian Mixture.

Furthermore, the GGM-SSM does not need one-to-one point correspondences but relies solely on point correspondence probabilities for the computation of mean shape and variation modes. Therefore, elaborate preprocessing of the observations in the data set to establish correspondences becomes obsolete, no questionable correspondences between point clouds representing surfaces are assumed, and the number of points in the observation shapes may vary. The approach can be used for non-spherical surfaces and can be adapted to applications on data sets with different topologies as the connectivity between points does not play a role.

At the moment, all points of the observations are equally included into the computation of the model. However, the corresponding matrix computed by the EM-ICP registration contains information about the probability for each point of an observation to correspond to any of the points of the model. For future applications, a weighting of the influence of observation points on the final result might be interesting, e.g. in order to reduce the influence of outliers. The same applies to point sets which are not evenly distributed over the estimated surface. In that case, regions containing relatively many points exert a higher amount of impact on the computation of the registration transformation than regions with fewer points. This behaviour is very helpful when shape details should be modeled but for other cases it might not be desirable and could be balanced by assigning a weight to each point. A main advantage of working with point-based shape representation is the simplicity of the resulting model with respect to its power. In the literature however, rather surface-based models are applied as the surface offers additional information about the boundary of the shape. Here it has to be kept in mind that the quality of the surface information they use depends on image quality and on the segmentation method. In order to expose advantages and limits of the new model compared to state-of-the-art models, its performance has to be compared to other statistical shape models for different kinds of application. An elaborate evaluation is performed in chapter 4.

Evaluation of the GGM-SSM

In this chapter, the GGM-SSM method is submitted to an extensive evaluation. The aim is to quantitatively compare its performance to other SSM methods in the literature and to gather knowledge about its behaviour and characteristics for different types of shapes. In section 4.1, the performance measures which are commonly used to assess the quality of SSMs are presented and discussed, and several distance metrics that are suited for point-based SSMs are introduced. Following that, the performances of the GGM-SSM and a classical ASM method for unstructured point sets are compared on different synthetic and real training data in section 4.2. Section 4.3 is dedicated to an evaluation of the GGM-SSM in comparison to a MDL-based approach. In section 4.4 it is demonstrated on a real data example how the GGM-SSM can be used for automatic shape classification. This chapter is concluded with a critical consideration of the advantages and weaknesses of the developed model (section 4.5).

4.1 Performance Measures

4.1.1 Assessing SSM Quality

In order to assess the quality of a given statistical shape model, an objective performance measure is needed. The measures introduced in the PhD thesis of R.H. Davies in 2002 have become a common standard in the community [Davies 2002b, Styner 2003c, Heimann 2005]. A good SSM is expected to

1. be able to model formerly unseen shapes of the same shape class.

2. only deform to plausible shapes when deformed in the shape space spanned by the variation modes and constrained by the standard deviations.

The first requirement is called *generalization ability*. The generalization ability indicates how well a SSM is able to match new - that is unknown - shapes. This is important e.g. when using the SSM to segmentation problems. The generalization ability is tested in a series of leave-one-out experiments where it is analysed how closely the SSM matches an unseen observation. This is done in two steps: First, the optimal affine transformation is computed to align the shapes in space. Secondly, the optimal deformation coefficients are determined and used to deform the aligned SSM in order to optimize the matching. Finally, the distance of the deformed SSM to the left-out observation is measured.

The second requirement is called *specificity*. The specificity indicates if the modeled variability in the SSM actually is a variability found in the training data set. In other words, the model should not be able to generate illegal shapes. For estimating the specificity, a high number of random shapes has to be generated by submitting the mean shape of the SSM to random deformations in the shape space spanned by the variation modes. Therefore, random deformation coefficients are generated under a uniform distribution with zero mean and variances equal to the squared standard deviation of the respective SSM. Then, the distance of the random shapes to the respective most similar observation in the training data set is measured.

In practice, these performance measures quantify the quality of a SSM in terms of correspondence evaluation. This sometimes poses a problem for several reasons: First, usually no ground-truth shape correspondences are availabe for medical image objects. Secondly, the measures depend on the point distribution on the shapes. Due to different SSM methods, the points representing the final SSMs will not be positioned at the same locations. Therefore, the variability model will not capture the same shape variations. This problem is amplified when comparing SSMs based on different numbers of points as a SSM with a greater number of points is naturally able to model more variation. These and other shortcomings of the performance measures were recently addressed in the work of Ericsson and Karlsson who propose manually set ground-truth correspondence measures [Ericsson 2007] in an attempt to remedy the problems. They generate synthetic examples which demonstrate clearly that better performance measures do not necessarily mean better SSM. Especially for cases where one SSM models more variability - e.g. on a higher detail level - than a second SSM, the specificity measure does not reflect the better quality of the first SSM.

To exemplify, let us regard a data set where some of the observations feature a nose-like shape and other do not (figure 4.1(a)). Let us assume that SSM 1 is able to capture this detail in one of its variation modes but SSM 2 fails to do so (figure 4.1(b,c)). During the test series for specificity, SSM 1 will probably produce several shapes with noses (e.g. shown in figure 4.1(d)) - as these exist in the shape space spanned by its variation modes - whereas SSM 2 will not. Instead, SSM 2 will produce shapes with less variability (e.g. shown in figure 4.1(e)). Naturally, the distances of the deformed mean shapes with prominent shape details to the observations in the training data set are greater than those of the shapes generated by SSM 2 as illustrated in figure 4.1(f,g). Therefore, we deem the performance measure 'specificity' to be not very well suited for measuring the quality of a SSM regarding shape details which do not occur in all observations. Generally, it has to be kept in mind that the realistic quality of a SSM always depends on its field of application. For example, a SSM that is very well suited for segmentation tasks does not necessarily perform well in classification tasks.

In the following experiments, the generalization ability and - for the sake of completeness - also the specificity measures are evaluated.

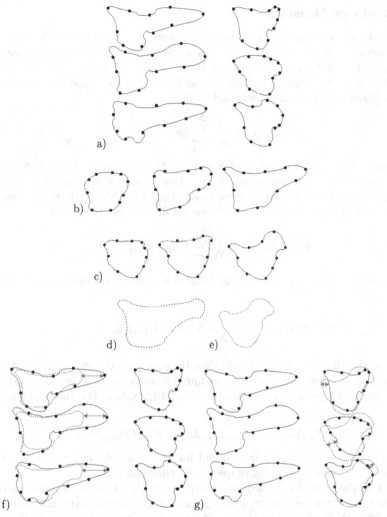

Figure 4.1: *Incoherent specificity example in 2D. a) Some observation examples of the training data set. b) SSM 1, the variability of the prominent feature in the training data set is captured. c) SSM 2 fails to capture the prominent feature in the training data set. d) Deformed mean shape in shape space spanned by the variation modes of SSM 1. e) Deformed mean shape in shape space spanned by the variation modes of SSM 2. f) Distance of deformed mean shape of SSM 1 to observations in training data set is measured. The Hausdorff distance is great due to the prominent feature. g) Distance of deformed mean shape of SSM 1 to observations in training data set is measured. The Hausdorff distance is smaller than the one of SSM 1.*

4.1.2   Distance Measures

A metric suited to evaluate the performance measures of a SSM obviously depends on
the representation of the shapes. As in this work the shape surfaces are represented
by point clouds, the distances are computed based on point coordinates. In order to
quantify the distance between two shapes S and M, an intuitive measure is the averaged
Euclidian distance between all corresponding points:

$$d_{CP}^2(S, M) = \frac{1}{N_S} \sum_{i=1}^{N_S} \|s_i - m_i\|^2$$

with N_S being the number of points of S and M. However, in the GGM-SSM no one-
to-one correspondences are computed. Hence, the distance d from an observation S_k
with N_k points s_{ki} to the deformed mean shape M_{def} with N_m points m_j is defined as
the square root of the normalized sum of squared differences (SSD) with

$$d^2(S_k, M_{def}) = \frac{1}{N_k} \sum_{i=1}^{N_k} \|s_{ki} - m_{ki}\|^2$$

where $m_{ki} = \arg\min_{m_j} \|s_{ki} - m_j\|$. This distance measure is not symmetric, hence, we
also compute

$$d^2(M_{def}, S_k) = \frac{1}{N_m} \sum_{j=1}^{N_m} \|s_{kj} - m_j\|^2$$

where $s_{kj} = \arg\min_{s_{ki}} \|s_{ki} - m_j\|$. In addition, the maximum distance $d_{max}(S_k, M_{def})$ is
computed as the maximal minimal distance found from S_k to M_{def} for $\|s_{ki} - m_{ki}\|$ with
$m_{ki} = \arg\min_{m_j} \|s_{ki} - m_j\|$ and respectively $d_{max}(M_{def}, S_k)$. The Hausdorff distance
is then

$$H(S_k, M_{def}) = max\left(d_{max}(S_k, M_{def}), d_{max}(M_{def}, S_k)\right).$$

This symmetric measure is especially useful for evaluating SSMs on data sets where
some observations feature different shape details than others.

Obviously, the measures defined above depend on the closeness of points after the
fitting which does not necessarily always represent the actual shape similarity. For
example, different distributions of landmarks over the estimated surface of the observations might affect the results. A more independent method would be to measure
the volume overlaps between the fitted shapes. However, as the GGM-SSM is based
on unstructured point sets, a binary representation can only be approximated for each
shape. This is done when comparing the GGM performance to the performance of an
MDL-based SSM in section 4.3. Here, the Jaccard coefficient is used to compute the
symmetric overlap of shape volumes A and B:

$$C_T = \frac{|A \cap B|}{|A \cup B|}.$$

It has to be kept in mind however that the Jaccard coefficient does not reflect well if shape details - which do not contribute much to the overall volume - are modeled or not.

For computing the distances between a SSM and a given observation, first the mean shape of the SSM is aligned with the observation. Then, the optimal deformation coefficients have to be computed. For the GGM-SSM, this is done by optimizing equation (3.13) with respect to the deformation coefficients ω_p. Here, $k = 1$ and S_1 equals the observation in question. The resulting coefficients are used to deform the aligned SSM in order to optimize the matching. Finally, the distance of the deformed SSM to the observation is measured.

4.2 Comparison to an ICP-SSM

In this section the performance of the GGM-SSM is evaluated in comparion with another SSM which is also based on unstructured point sets. As opposed to the GGM-SSM, the henceforward called *ICP-SSM* relies on one-to-one correspondences. It is based on the classical ASM approach applied to unstructured point sets represented by varied numbers of points. The ICP-SSM is computed as follows:

1. The observations in the training data set are aligned with an initial mean shape employing affine Iterative Closest Points (ICP) registrations. (For the algorithm see section 2.2.1.) The ICP matches the observations and determines correspondences simultaneously. The correspondences are explicitly given by the nearest neighbour for each point.

2. The mean shape is computed on the aligned observations. Registration and mean shape computation are iterated. For the data sets used in practice we found that after 2 or 3 iterations, the mean shape does not change significantly anymore.

3. A principal component analysis is performed on the aligned data set to determine the eigenmodes and the eigenvalues. Here, a SVD is applied on the covariance matrix cleared of the mean.

The computation of a distance between ICP-SSM and a given observation follows the same procedure as explained for the GGM-SSM in section 4.1.2. Here, the deformation coefficients ω_p are computed by solving the linear system of equation (2.2) where M equals the observation in question.

The performances of the two SSM computations are evaluated on three different synthetic data sets in sections 4.2.1 and 4.3 and on a real data set containing brain structures in section 4.2.2.

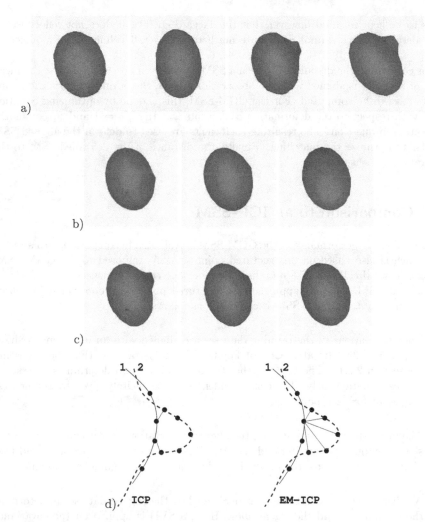

Figure 4.2: *a) Observation examples of a synthetic training data set featuring two distinctive shape classes (ellipsoids with bump and ellipsoids without bumps). b,c) Results of a SSM built on exact correspondences (ICP-SSM)(b) and of a SSM built on correspondence probabilities (GGM-SSM)(c) for the training data. For both SSMs, the mean shape (middle), and the mean shape deformed with respect to the first eigenmode $(\bar{M} - 3\lambda_1 \vec{v}_1$ (left) and $\bar{M} + 3\lambda_1 \vec{v}_1$ (right)) are depicted. d) One-to-one correspondence versus correspondence probabilities. Left: ICP registration, each point on contour 1 corresponds to the closest point on contour 2. Right: EM-ICP registration, each point on contour 1 corresponds with a certain probability to all points on contour 2.*

Table 4.1: *Ellipsoid shape results. Shape distances found in generalization experiments (leave-one-out tests) with ICP-SSM approach and with GGM-SSM approach. The distances and associated standard deviations are given in cm.*

	ICP-SSM	GGM-SSM
mean distance target to source	0.207 ± 0.048	0.139 ± 0.032
mean distance source to target	0.214 ± 0.058	0.125 ± 0.030
maximal distance target to source	0.431 ± 0.036	0.415 ± 0.042
maximal distance source to target	0.567 ± 0.186	0.380 ± 0.044

Table 4.2: *Ellipsoid shape specificity results on 100 random shapes found with ICP-SSM approach and with GGM-SSM approach. The average distance from the randomly deformed mean to the respective closest observation is measured. The distances and associated standard deviations are given in cm.*

	ICP-SSM	GGM-SSM
average distance	0.102 ± 0.003	0.160 ± 0.022

4.2.1 Synthetic Data

4.2.1.1 Ellipsoids

The determination of correspondences between unstructured point sets is especially difficult when one shape features a certain structure detail and the other one does not. For an experimental evaluation, a training data set is generated containing two distinctive shape classes. The data set consisted of 9 ellipsoids featuring a bump and 9 ellipsoids without bump. Their sizes as well as the bump sizes and their 3D rotations in space varied. For several observation examples, see figure 4.2(a). The long axes measure around $70mm$. The observations are represented by $276 - 337$ points respectively, and the point distances average $0.24mm$. The GGM-SSM as well as the ICP-SSM are computed for these data. For the computation of the GGM-SSM, the following parameters were chosen: $\sigma_{start} = 0.5mm$, reduction factor $= 0.7$, 7 iterations (EM-ICP multi-scaling) with 15 SSM iterations. For the ICP-SSM, the ICP is iterated 40 times. Then the tests for generalization ability are performed in a series of leave-one-out experiments. The specificity for both models was tested using 100 randomly generated shapes.

Results: The respective mean shapes and deformations according to the first mode of variation for the GGM-SSM as well as the ICP-SSM are illustrated in figure 4.2(b,c). Clearly, the GGM-SSM models the bump of the ellipsoids in its first mode of variation while the ICP-SSM fails to do so. Quantitatively, this is backed up by the results obtained in the evaluation of the performance measures. The values of the generalization ability are depicted in table 4.1 for both SSMs. The mean distances of the left-out observation to the respective fitted SSM are about 35% smaller for the

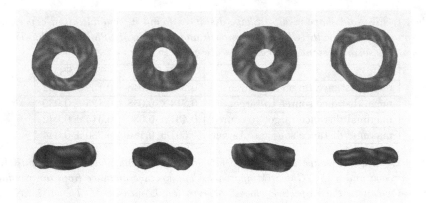

Figure 4.3: *Four observation examples of a synthetic training data set featuring bagel shapes, shown from above and from the side.*

GGM-SSM ($0.139cm$ and $0.125cm$) than for the ICP-SSM ($0.207cm$ and $0.214cm$). Also the comparatively great Hausdorff distances indicate that the ICP-SSM is not able to successfully model the bump on the ellipsoid shapes.

The results for the specificity are depicted in table 4.2. The average distances of the randomly deformed GGM-SSM mean shape to the respectively closest observation in the training data set are a bit higher than the average distances of the ICP-SSM. As a visual inspection as well as the generalization ability values strongly indicate the superior performance of the GGM-SSM on the given data, these specificity results corroborate the problems concerning the specificity measure as discussed in section 4.1.1.

The GGM-SSM based on the EM-ICP models the whole data set, it is able to represent the ellipsoids featuring a bump and those without as that deformation information is included in its variability model. The SSM based on the ICP however is not able to model the bump. This is due to the fact that the ICP only takes into account the closest point when searching for correspondence. Thus, the points on top of the bump are not necessarily involved in the registration process and do not contribute to the variability model. The EM-ICP, on the other hand, analyzes the correspondence probability of *all* points, therefore, also the points on top of the bump are taken into account. These two concepts are illustrated in figure 4.2(d).

4.2.1.2 Bagel Shapes

Another interesting problem regarding statistical shape models are shapes featuring non-spherical surfaces. Here, the aim is to evaluate the performance of the GGM-SSM on shapes with genus 1 topology. In the case of a simple ring torus, the surface can be created in Euclidean space by revolving a circle about an axis in its plane. Non-spherical shapes cannot be modeled by all current SSM computation methods, e.g. the SPHARM and the MDL approaches (section 2) work exclusively for spherical topologies.

For the generation of the data set, the rotation axes did not necessarily lie in a plane. Furthermore, the inner and outer radii from observation to observation are varied which means that our bagel shapes are not radially symmetric. For some observation examples see figure 4.3. A synthetic data set was generated containing 15 observations. The observations are represented by $332 - 512$ points, their bounding boxes measure about $1500 \times 1500 \times 500 mm^3$ and the point distances average $82mm$. The GGM-SSM as well as the ICP-SSM are computed for these data. For the computation of the GGM-SSM, the following parameters were chosen: $\sigma_{start} = 100mm$, reduction factor $= 0.9$, 5 iterations (EM-ICP multi-scaling) with 15 SSM iterations. Then the tests for generalization ability were performed in a series of leave-one-out experiments. The specificity for both models was tested using 500 randomly generated shapes.

Results: The mean shape as well as the deformations according to the first two variation modes of GGM-SSM and ICP-SSM are displayed in figure 4.4. As can be seen, the first variation mode principally models the thickness of the bagel while the second variation mode mainly model its flexion. The quantitative evaluation results for the generalization ability are shown in table 4.3. The values show a better generalization ability for the GGM-SSM than for the ICP-SSM as the mean distances are more than 30% smaller. The Hausdorff distances show that apparently the GGM-SSM ($75.04mm$) captured more shape variation than the ICP-SSM ($109.05mm$). An illustration is shown in figure 4.5. The flexion in the bagels seems to lead to erroneous correspondences in the ICP-SSM. Looking closer at the leave-one-out series, it could be established that especially the bagel shapes of which the axes do not lie in planes are matched better by the GGM-SSM. This is illustrated in figure 4.6 with an example. The results for the specificity evaluation are depicted in table 4.4. The specificity values are a little better for the GGM-SSM than for the ICP-SSM.

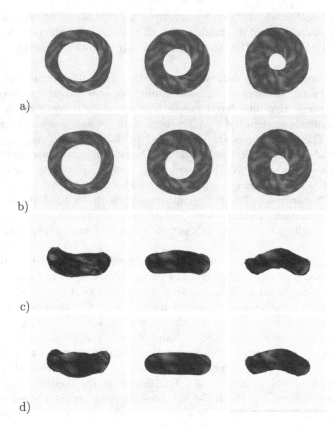

Figure 4.4: *SSM results for bagel data set. GGM-SSM (a,c) and ICP-SSM (b,d) deformations to first (a,b)and second (b,c) variation mode: Mean shape (middle), and mean shape deformed according to variation modes, left: $\bar{M} - 3\lambda_p \vec{v}_p$ and right: $\bar{M} + 3\lambda_p \vec{v}_p$.*

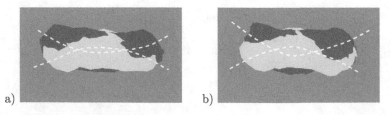

Figure 4.5: *Schematic illustration of modeled amount of flexion. Deformations according to second variation mode for ICP-SSM (a) and GGM-SSM (b). A higher amount of flexion seems to be modeled by the GGM-SSM.*

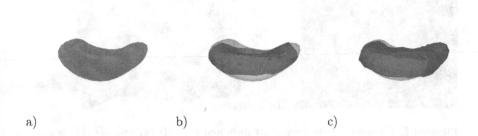

a)   b)   c)

Figure 4.6: *Generalization ability example for one left-out observation with high amount of flexion. a) Left-out observation featuring high amount of flexion. b) Fitting result of ICP-SSM. c) Fitting result of GGM-SSM. In b) and c), the left-out observation is coloured in light grey with low opacity, the results of ICP-SSM and GGM-SSM are coloured in dark grey.*

Table 4.3: *Torus shape generalization results. Shape distances found in generalization experiments with ICP-SSM approach and with GGM-SSM approach. The distances and associated standard deviations are given in mm.*

	ICP-SSM	GGM-SSM
mean distance target to source	41.47 ± 6.42	31.08 ± 15.01
mean distance source to target	38.25 ± 5.18	29.34 ± 12.68
maximal distance target to source	87.73 ± 11.10	77.83 ± 31.09
maximal distance source to target	109.05 ± 35.14	75.04 ± 25.36

Table 4.4: *Torus shape specificity results on 500 random shapes found with ICP-SSM approach and with GGM-SSM approach. The distances and associated standard deviations are given in mm.*

	ICP-SSM	GGM-SSM
average distance	45.95 ± 2.52	33.82 ± 5.47

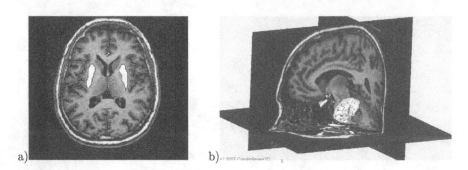

Figure 4.7: *CT-images with segmented putamen in a 2D (a) and 3D (b) view.*

4.2.2   Brain Structure MR: Putamen

In this section, the performance of the GGM-SSM on brain structure data is evaluated. The data has been collected in the framework of a study on hand dystonia and the possible influence of this disease on the shape of the putamen, a structure belonging to the basal ganglia situated close to the caudate nucleus. The MR images as well as the segmentations of the putamen were kindly provided by the Hôpital La Pitié-Salpêtrière, Paris, France. An example of left and right putamen is shown in figure 4.7. The MR images contain $255 \times 255 \times 105$ voxels of size $0.94mm \times 0.94mm \times 1.50mm$. The training data set for this experiment consists of $N = 20$ left segmented putamens (approximately of size $20mm \times 20mm \times 40mm$) which are represented by min 994 and max 1673 point. Some observation examples are shown in figure 4.8(a). The computation of a SSM for the putamen data might be useful either for segmentation purposes or for an analysis of the shape variability in patient and control groups.

The GGM-SSM as well as the ICP-SSM are computed for these data and then tested for generalization ability in a series of leave-one-out experiments. The specificity for both models was tested using 500 randomly generated shapes. For the computation of the GGM-SSM, the following parameters were chosen: $\sigma_{start} = 4mm$, reduction factor $= 0.85$, 10 iterations (EM-ICP multi-scaling) with 5 SSM iterations. For the ICP-SSM, the ICP is iterated 50 times. Most of the parameter values were found in an heuristic way.

Results: The resulting mean shapes and deformations according to the first two variation modes are shown in figure 4.8(b,c) for the GGM-SSM and in figure 4.8(d,e) for the ICP-SSM. The mean shapes of both approaches resemble. However, the first and second variation mode of the GGM-SSM model more shape details than the first and second eigenmodes of the ICP-SSM. This visual impression is confirmed by the values found for the generalization ability as depicted in table 4.5. The generalization ability is computed in dependence of the number n of variation modes used. The results for the first $n = 5$, $n = 10$ and $n = 18$ variation modes are shown.

Table 4.5: *Shape distances found in generalization experiments with the ICP-SSM approach and with GGM-SSM approach. The generalization ability was tested for the first $n = 5$, $n = 10$ and $n = 18$ variation modes. The distances and associated standard deviations are given in mm.*

	ICP-SSM	GGM-SSM
5 variation modes		
average mean distance + std dev. in mm	0.634 ± 0.090	0.512 ± 0.083
average maximal distance + std. dev. in mm	4.478 ± 0.927	2.929 ± 0.576
10 variation modes		
average mean distance + std. dev. in mm	0.623 ± 0.099	0.490 ± 0.088
average maximal distance + std. dev. in mm	4.449 ± 0.909	2.496 ± 0.445
18 variation modes		
average mean distance + std. dev. in mm	0.610 ± 0.089	0.471 ± 0.076
average maximal distance + std. dev. in mm	4.388 ± 0.930	2.559 ± 0.563

Table 4.6: *Shape distances found in specificity experiments (500 random shapes) with ICP-SSM approach and with GGM-SSM approach using 18 eigenmodes.*

	ICP-SSM	GGM-SSM
average mean distance + std. dev. in mm	0.515 ± 0.117	0.463 ± 0.052

Obviously, the number of variation modes controls the accuracy of the deformed SSM. The GGM-SSM performed better for all cases with a mean distance of 0.471 for the GGM-SSM and a mean distance of $0.610mm$ for the ICP-SSM under the use of 18 variation modes. It is interesting to see that the performance difference between the two SSMs increased a little with a higher number of variation modes. The mean distance decrease regarding the case of $n = 5$ variation modes and the case of $n = 18$ variation modes is about 5% using the SSM-ICP and about 8% using the GGM-SSM. Commonly, the variation modes with great standard deviations model the obvious variabilities as e.g. thickness or torsion in space while the variation modes with smaller standard deviations model the shape details. The Hausdorff distance in the GGM-SSM is more than 40% (nearly $2mm$) smaller than the Hausdorff distance of the ICP-SSM. This result again indicates that the GGM-SSM is better able to capture shape details than the ICP-SSM. The results for the specificity evaluation are depicted in table 4.6. The specificity values are a little better for the GGM-SSM than for the ICP-SSM.

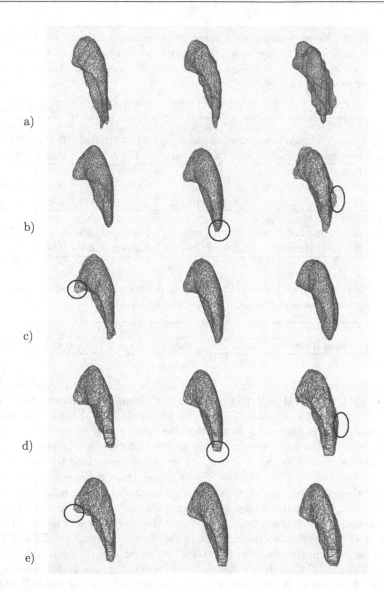

Figure 4.8: *Real training data set featuring the putamen. a): Observation examples. b)/c): GGM-SSM. d)/e): ICP-SSM. Mean shapes (middle) and mean shapes deformed with respect to the first (b,d) and second (c,e) variation mode. Left: $\bar{M} - 3\lambda v_{1,2}^{\rightarrow}$ and right:$\bar{M} + 3\lambda v_{1,2}^{\rightarrow}$. The regions in circles mark shape details which are represented by the GGM-SSM and which are not modeled by the ICP-SSM.*

4.3   Comparison to ICP-SSM and MDL-SSM

In this section, the performance of the GGM-SSM is evaluated in comparison to a SSM whose computation is based on the minimization of a Maximum-Description-Length (MDL). This SSM method is explained in detail in section 2.3.2. Basically, the MDL is used to optimize the distribution of corresponding points on the surfaces of the observations in the training data set. Here, the best point distributions or correspondences yield the best SSM in terms of simplicity. One key step in computing a MDL-SSM is the movement of points on the surfaces of the respective observations. Hence, as it needs explicit surface information, the MDL approach is not suited to compute a SSM for unstructured point sets. Nevertheless, an interesting prospect is to contrast the performance of the ICP-SSM and the GGM-SSM with a MDL-SSM to point out the differences in the approaches and to position our method in the state-of-the-art. In order to be able to use the MDL-method, a training data set of observations with surfaces represented by triangulated points has to be generated.

Data Set: Unlike the GGM-SSM, the MDL-method can only be applied for data with spherical topologies. The objective is to test both approaches as well as the ICP-SSM on non-convex shapes which can be challenging, e.g. as points lying close do not necessarily belong to the same part of the shape. Moreover, points with similar normal vector direction do not necessarily lie close to each other. A synthetic data set is generated containing 15 observations shaped like bananas, see figure 4.9. The observations are represented by triangulated meshes. In order to obtain meaningful results, the variability in the training data set is high: The curvature of the banana as well as the size, thickness and orientation in space change from observation to observation. The sizes of their bounding boxes measure around $480 \times 720 \times 260 mm^3$. The number of points range from minimum 386 points to maximum 642 points. The point distances average $29.3mm$.

Set-Up: The MDL-SSM experiments on this data were performed by Tobias Heimann of the German Cancer Research Center (Department of Medical and Biological Informatics) who kindly provided his evaluation results for this section. The alignment of observations is done using a generalized Procrustes analysis in similarity mode. The final number of points is set to 648.

For the computation of the GGM-SSM, the following parameters were chosen: $\sigma_{start} = 15 - 50mm$ (dependent on the observation shape), reduction factor $= 0.7 - 0.9$, 10 iterations (EM-ICP multi-scaling) with 5 SSM iterations. For the ICP-SSM, the ICP is iterated 50 times. Most of the parameter values were found in an heuristic way. The mean shapes of the GGM-SSM as well as of the ICP-SSM contain 446 points which is 200 points less than used by the MDL-SSM.

For determining the performance measures in these experiments, the average point distances as introduced in section 4.2 are only a well-suited metric when SSMs with

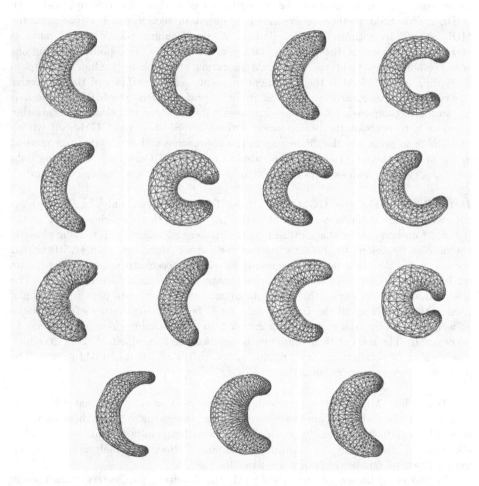

Figure 4.9: *Synthetic training data set: Non-convex banana shapes with 15 observations represented by triangulated meshes.*

equal numbers of points and similar point distributions are compared. This is not the case when comparing the MDL-SSM to the GGM-SSM as the MDL method moves the points over the surfaces and can add any number of points. Therefore, in the experiments the Jaccard coefficient (or Tanimoto coefficient) is used as distance metric instead of the point distances. To do so, a binary representation has to be approximated for all observations as well as for each deformed SSM. For the GGM-SSM a well as the ICP-SSM this is done by keeping the edges of the triangles in the initial mean shape for the representation of the final mean shape and its deformations. As the GGM-SSM is based on unstructured point sets, this procedure could theoretically lead to contorsions of the mesh but this was not the case in the experiments.

The generalization ability is evaluated in a series of leave-one-out tests. The distances were measured in dependence of the number n of employed variation modes ranging from $n = 0$ to $n = 13$. For the specificity, 500 random shapes are generated. Due to the high computational time when generating the binary volume representation, the alignment of each randomly deformed mean shape with all observations is omitted. Instead, all observations are aligned once with the undeformed mean shape. That way, for each randomly deformed mean shape, only one binary representation has to be computed and compared to the observations.

Results: The mean shapes and the deformations according to first, second and third mode of variation are depicted for the ICP-SSM and the GGM-SSM in figures 4.10 and 4.11. The first three variation modes roughly represent similar variabilities. However, it is noticeable that the GGM-SSM variability model is strongly focused on the region of the banana tips whereas the ICP-SSM rather models global variation of the banana shapes. The values resulting from the testing series of the generalization ability are illustrated in figure 4.12 for ICP-SSM, GGM-SSM and MDL-SSM methods. The volume overlap between left-out observation and fitted SSM is used as distance metric. Regarding these values, the experiments revealed that the MDL-SSM has a higher generalization ability with an average Jaccard coefficient of 0.92 than the GGM-SSM (Jaccard coefficient = 0.88) and the ICP-SSM (Jaccard coefficient = 0.86). As - contrary to point-based methods - the MDL-SSM method makes use of the observation surfaces as additional information, this result is not surprising. In particular, it has to be kept in mind that the MDL-SSM approach optimizes the distribution of corresponding points over the observation surfaces which is one of its great strengths. The GGM-SSM method however uses the initial point locations. Regarding the banana shapes, the point distribution at the banana tips is more dense than on the banana corpus. Using the GGM-SSM, this leads to a more detailed modeling of the banana tip regions. Unfortunately, a volume overlap metric does not necessarily reflect if shape details are well modeled.

Besides, the following bias in the MDL-SSM generalization ability values has to be considered: For SSMs where the correspondences are described by monotonous parameterization functions the parameterization of the left-out function is unknown.

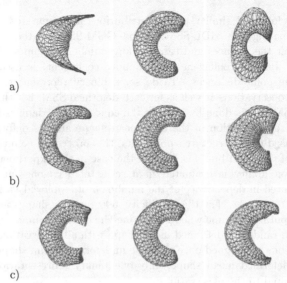

a)

b)

c)

Figure 4.10: *GGM-SSM for the banana shape data set. Mean shapes (middle) and mean shapes deformed according to the first (a), second (b) and third (c) variation mode.*

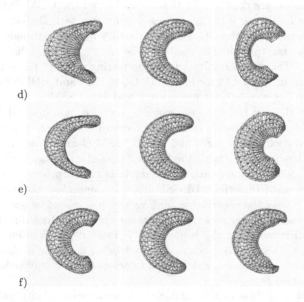

d)

e)

f)

Figure 4.11: *ICP-SSM for the banana shape data set. Mean shapes (middle) and mean shapes deformed according to the first (a), second (b) and third (c) variation mode.*

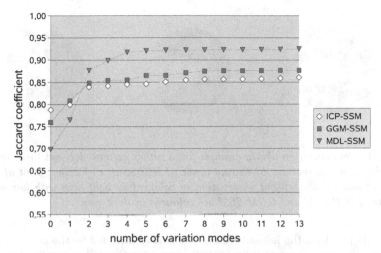

Figure 4.12: *Generalization ability. The generalization ability was tested in leave-one-out tests for the banana shapes. Here, the average overlap between deformed mean shape and left-out observation is presented for the MDL-SSM, the GGM-SSM and the ICP-SSM.*

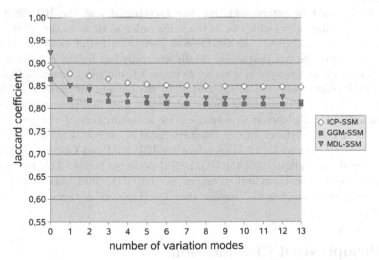

Figure 4.13: *Specificity. The specificity was tested for the banana shapes using 500 testing shapes. Here, the average overlap between randomly deformed mean shape and closest observation is presented for the MDL-SSM, the GGM-SSM and the ICP-SSM. The random deformation followed a natural distribution with σ equal to the standard deviations of the respective model.*

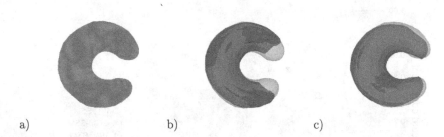

a)   b)   c)

Figure 4.14: *Generalization ability example for a rather extreme left-out torus observation. a) Left-out observation. b) Fitting result of ICP-SSM. c) Fitting result of GGM-SSM. In b) and c), the left-out observation is coloured in light grey with low opacity, the results of ICP-SSM and GGM-SSM are coloured in dark grey.*

To solve this problem, the left-out shape is normally included in the correspondence localisation. This procedure finally leads to an over-estimated generalization ability [Ericsson 2007].

The specificity values are illustrated in figure 4.13. Here, the GGM-SSM and the MDL-SSM obtained very similar overlap values while the ICP-SSM obtained values a little higher.

Overall, it could be established that the GGM-SSM and the ICP-SSM obtain generalization ability values which lie in the same order as those of the MDL-SSM for the given data set. Moreover, the GGM-SSM performed better than the ICP-SSM. This is again due to the fact that shape details are easily lost for the ICP-SSM. This is demonstrated with an example of a rather extreme left-out observation in figure 4.14. The ICP-SSM adapts very well to the corpus of the banana but fails to deform into its tip. Yet, the variability model of the GGM-SSM is able to represent the tip region of the banana. This behaviour is confirmed by an evaluation of the generalization ability under a point distance metric (as introduced in section 4.1.2 and as used for the experiments in section 4.2.1). The values for ICP-SSM and GGM-SSM which are depicted in table 4.7 indicate that the GGM-SSM performs better. This becomes clear especially regarding the Hausdorff distances as the GGM-SSM obtains a Hausdorff distance of $53,78mm$ which is 37% smaller than the Hausdorff distance of the ICP-SSM ($83,87mm$).

4.4   Unsupervised Classification

In this section the GGM-SSM is applied to a classification problem. This can be done directly by exploiting the observation parameters computed during the GGM-SSM computation. Here, the final deformation coefficients ω_{kp} represent the amount of variation for the respective observation S_k according to each variation mode v_p. Therefore, in-

Table 4.7: *Banana shape generalization results. Shape distances found in generalization experiments with ICP-SSM approach and with GGM-SSM approach. The distances and associated standard deviations are given in mm.*

	ICP-SSM	GGM-SSM
mean distance target to source in *mm*	15.75 ± 2.28	16.48 ± 3.24
mean distance source to target in *mm*	26.35 ± 12.78	17.81 ± 2.75
maximal distance target to source in *mm*	36.23 ± 4.60	53.78 ± 7.33
maximal distance source to target in *mm*	83.87 ± 54.58	43.81 ± 8.41

formation about shape characteristics can be gained by evaluating the deformation coefficients [Hufnagel 2007b]. In SSM methods where the deformation coefficients are not computed during optimization of the model, their determination is less straightforward.

In an experimental evaluation, the deformation coefficients directly serve as a classification measure regarding the shape of the observations S_k. To do so, feature vectors $\omega_k = (\omega_{k1}, \omega_{k2}, ..., \omega_{kn})$ are formed and then used as input for a k-means clustering. This approach is tested on the synthetic data set of ellipsoids as used in section 4.2.1.1. The data set consists of two shape classes as it contains ellipsoids with and without 'bump' as can be seen exemplarily in figure 4.2(a). An average Rand index [Rand 1971] of 0.95 is employed for the k-means clustering. The resulting two classes coincide with the 'bump' and 'without bump' classes, see figure 4.15 for an example of the values of the 2D feature vectors $(\omega_{k1}, \omega_{k2})$.

Tame approach is applied to classify the putamen data set as presented in section 4.2.2. As the data was gathered in a study about hand dystonia, a relation of shape and disease might exist. In order to analyse the shapes, the data is tested for statistically significant shape differences between dystonia patients and control group after affine normalizations. Again feature vectors $\omega_k = (\omega_{k1}, \omega_{k2}, ..., \omega_{kn})$ are formed and used as input for a k-means clustering. In this case, no two distinct shape classes were found (see figure 4.16 for the values of the 2D feature vectors $(\omega_{k1}, \omega_{k2})$). This confirms the presumption of the concerned physicians.

4.5 Discussion

An accurate and robust modeling of variability is an important feature of a SSM, particularly when it is employed to the segmentation of anatomical structures for radiotherapy or surgery planning where the precision must be high. In order to learn about the qualities of the GGM-SSM as well as its standing in the state-of-the-art, the evaluation has been divided into two experiments: The first part was aimed at an analysis of the GGM-SSM performance in comparison to another SSM for unstructured point sets (ICP-SSM). The second part of the evaluation investigated the GGM-SSM

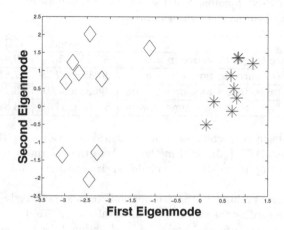

Figure 4.15: *2D deformation coefficient feature vectors $(\omega_{k1}, \omega_{k2})$ for the first two eigenmodes of the ellipsoid data set. Observations 'with bump' are represented by diamonds, observation 'without bump' are represented by stars.*

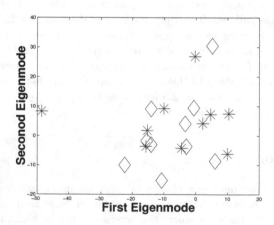

Figure 4.16: *2D deformation coefficient feature vectors $(\omega_{k1}, \omega_{k2})$ for the first two eigenmodes of the putamen data set. 'Control' observations are represented as diamonds and 'patient' as stars.*

performance in comparison to a well established method which uses surface information (MDL-SSM).

A principal difference between the ICP-SSM and the GGM-SSM is the interpretation of correspondence. While the ICP-SSM is based on one-to-one point correspondences, the GGM-SSM implements a probabilistic correspondence concept which allows to take into account all points of all shapes. This is advantageous on the one hand as all shape details are integrated into the variability model. On the other hand, the approach is less sensitive to possible outliers. By evaluating the generalization ability values of GGM-SSM and ICP-SSM for the synthetic data set of ellipsoid shapes, it could be established that shape details which are not captured very well by the ICP-SSM are effectively captured and modeled by the GGM-SSM. This is especially the case for training data where not all observations feature the same shape details. Furthermore, when testing both SSMs on shape data with a global variation in its flexion angle, the generalization ability values indicate that the ICP-SSM did not model well the variability of flexion. The performance measures of GGM-SSM and ICP-SSM in the experimental evaluation on real brain data show a similar picture. The GGM-SSM is better able to capture shape details which can be observed by a visual inspection of the principal variations modeled by the variability models and which is also reflected in the generalization ability values. Still, the ICP-SSM faster and easier to handle than the GGM-SSM as less parameters have to be estimated beforehand. The relatively high computational time of the GGM-SSM is mainly due to the costly update of variation modes which involves several matrix multiplications with matrices $\in \mathbb{R}^{3N_m \times n}$ with number of mean shape points N_m and number of variation modes n. However, the analysis of shape in medical practice is generally no time sensitive matter.

As argued in section 4.1.1, we doubt the meaningfulness of specificity values regarding the quality of a SSM. These doubts were confirmed by the results obtained for the SSMs in the ellipsoid data set. Here, the generalization ability as well as visual inspection clearly indicate a superior performance of the GGM-SSM, but still the ICP-SSM obtain better specificity values.

The second part of the evaluation serves to position the GGM-SSM in the state-of-the-art by outlining its advantages and weaknesses compared to the well-accepted surface-based MDL-SSM method. The MDL-SSM approach makes use of surface information for the modeling of the training data set. During SSM computation, points are added and moved over the observation surfaces in order to find optimal correspondences. Therefore, the MDL-SSM is more flexible than the GGM-SSM as the results do not depend on the original point distribution in the observation meshes. Yet, it has to be kept in mind that the MDL-SSM is explicitly defined on surface representations for spherical topologies. Hence, it cannot be employed for the evaluation on the bagel shape training data but a training data set with banana-shaped observations was designed. As the training data set contains observations with very non-convex shapes, we deem the obtained results of the MDL-SSM as well as the GGM-SSM to be quite good. In the generalization ability experiments, the MDL-SSM performed better than the GGM-SSM by obtaining a Jaccard coefficient which is 3.4% greater than the GGM-SSM and 6.4% greater than the ICP-SSM. The difference between MDL-SSM and GGM-SSM in the volume overlaps is clearly visible but small enough to suggest the right of existence for the GGM-SSM, especially considering that the usage of surfaces is arguable for the reasons formulated in section 1. Moreover, the left-out observations in the experiment series for the generalization ability of the MDL-SSM method have been part of the correspondence localisation step, thus, the values of the generalization ability might be over-estimated. The analysis of the generalization ability for the banana training data set measured by point distance metrics shows that the GGM-SSM outperforms the ICP-SSM; the ICP-SSM fails to model shapes featuring a rather extreme convexity.

In order to compute a GGM-SSM of high quality, particular attention has to be paid to the choice of parameters in the EM-ICP registration which have to be adapted to the problem at hand. As demonstrated in section 3.2.3, good results are obtained for a final standard deviation which lies in the same range as the average point distances in the observations. A reasonable choice for the reduction factor seems to lie between 0.7 and 0.9 which led to good results in the experiments performed in the framework of this thesis. The number of GGM-SSM iterations is kept as small as possible to reduce computational cost.

From the evaluation results, it can be concluded that the GGM-SSM method is capable to model different kinds of shapes with high precision. Due to the probabilistic modeling of correspondence, the GGM-SSM outperforms the ICP-SSM for observations with irregular shape differences. The GGM-SSM does not need surface information and is well suited to model non-spherical topologies as well as coupled structures in one unified variability model. Therefore, the GGM-SSM is fit for shape analysis of various types of anatomies which makes it very flexible regarding potential application domains.

CHAPTER 5

Using the GGM-SSM as a Prior for Segmentation

Segmentation algorithms play a major role in medical image analysis. However, due to typical medical image characteristics as poor contrasts, grey value inhomogeneities, contour gaps, and noise the automatic segmentation of many anatomical structures remains a challenge. Low-level algorithms as region growing, thresholding or simple edge-detection are often bound to fail or require heavy user interaction to lead to acceptable segmentation results in 3D images. In order to overcome these problems, a very popular approach is to employ models which incorporate a priori knowledge about mean and variance of shape or grey levels of the structure of interest. These models serve to constrain the resulting segmentation contour to probable shapes as defined by the underlying training data set. The concept of shape priors in segmentation methods has been analysed in section 2.4.

In this chapter, a framework is developed for the integration of the GGM-SSM created in chapter 3 as a shape prior for kidney segmentation. In this new method, prior shape knowledge represented by the GGM-SSM is combined with prior information about typical grey value intensity distributions inside and outside the organ to be segmented. The chapter is structured as follows: First an overview is given about the employment of intensity distribution knowledge in medical image segmentation, and the initial placement problem is explained in section 5.1. In section 5.2, a sound mathematical framework is developed which integrates the GGM-SSM into an implicit level set scheme, and the method is evaluated on the segmentation of the kidney from CT images. In section 5.4, the level set framework is extended to multiple-object segmentation, and the algorithm is applied to hip joint segmentation. The chapter is concluded with section 5.5 where the approach of combining an explicitly represented SSM and an implicitly represented segmentation contour is discussed.

5.1 Initialization

5.1.1 Distribution Models for Prior Intensity Knowledge

Beside the prior knowledge about the shape, knowledge-based segmentation methods often integrate information about the grey value appearance of the organ which are extracted from a training data set. Classical segmentation techniques using SSMs mostly rely on edge-detection [Cootes 1992, Székely 1996, Staib 1996, Wang 2000]. Recent methods propose the utilization of a priori knowledge about intensity information on its own [Nain 2007, Andreopoulos 2008] or in combination with boundary detection [Huang 2004] in order to exploit available image information which generally leads to methods that are more robust and effective.

In point-based SSMs, a widely-used method is to generate local appearance models. The first local appearance model was presented by Cootes et al. [Cootes 1993] who proposed to sample intensity information around each landmark in normal direction. This is done for all observations in the training data set in order to determine mean value and principal modes of variation of grey value appearance over the corresponding landmarks. During segmentation, the intensity model profiles of each SSM landmark are compared to the current point profile samples of the deformed SSM in the image in order to optimize the fit. The local appearance models range from simple Gaussian intensity profile models and Gaussian gradient profile models [Cootes 1994] to non-linear intensity profile models [de Brujine 2002] and histogram region models [Brunelli 2001, Freedman 2005].

A local appearance model as described here is not immediately usable for our GGM-SSM as one-to-one correspondences over the observations are needed in order to extract statistical knowledge about the grey values at one specific point of the model. Therefore, a global appearance model is employed which means that a priori knowledge about the intensity distributions in the regions inside and outside the organ has to be extracted. In general, an intensity distribution model consists of two probability density functions which model the occurrence of grey values inside (p_{in}) and outside (p_{out}) the organ. A straightforward method is to sample the grey values of organ pixels x in the training data set and compute a mean grey value μ as well as a standard deviation σ_g. Then the probability of a voxel grey value $g(x)$ to occur inside the organ is estimated with $p_{in}(g) = \frac{1}{\sqrt{2\pi}\sigma_g} \exp(-\frac{(\mu-g)^2}{2\sigma_g})$. Then, $p_{out}(g) = 1 - p_{in}(g)$ could directly estimate the probability of a voxel grey value $g(x)$ to occur outside the organ. However, for most soft tissue organs neither the organ tissue nor the surrounding tissue belong to only one tissue class and additionally, noise has to be taken into account. Therefore, a classification using a mixture of Gaussians should lead to a more reliable model of intensity distributions. Thus, we take advantage of a pattern classification technique introduced by Duda and Hart [Duda 1973] which is based on the so-called kernel density approximation to estimate the point distribution function of a random variable. This non-parametric method was first proposed by Parzen [Parzen 1962] in order to solve problems in the field of time series analysis. In short, the method works as follows: For a given random sample $X = \{x_1, ..., x_n\}$ the value of the underlying but unknown probability density function $p(x)$ is sought. Using a kernel or window function $\varphi : \mathbb{R}^d \to \mathbb{R}$ with the properties $\varphi(u) > 0$ and $\int \varphi(u)du = 1$, it can be approximated

$$\hat{p}(x) = \frac{1}{n} \sum_{i=1}^{n} \frac{1}{h^d} \varphi\left(\frac{x - x_i}{h}\right).$$

The parameter h defines the width of the window and is generally chosen with respect to the size of the sample. A widely-used example for the window function is the Gaussian kernel $\varphi_{gauss}(x) = \frac{1}{\sqrt{2\pi}} \exp(-\frac{1}{2}x^2)$. The choice of window function φ and width h determines the smoothing effect on the estimated probability density function. In order to estimate the grey value density distributions for the inside of an organ as well as for its background, the intensities G_{in} and G_{out} are sampled around the surface of the organ:

$$G_{in} = \{g(x)|x \text{ inside organ and close to boundary}\}$$
$$G_{out} = \{g(x)|x \text{ outside organ and close to boundary}\}$$

In order to avoid the influence of to partial volume effects and segmentation inaccuracies, the sampling is done at a certain distance from the original organ boundary [Schmidt-Richberg 2009]. For an example of the sampling and the resulting grey value density distributions see figure 5.1.

5.1.2 Initial Placement Problem

The initial placement of any template in the image plays an important role regarding the quality of the segmentation result. Therefore, the initial location, transformation and deformation of the GGM-SSM has to be determined carefully. A position too far away from the organ region or an initial deformation too different from the organ shape in the image augments the risk of finding a local minimum which is not consistent with an acceptable segmentation. Aside from manual intervention which yields good results but is time-consuming [de Bruijne 2003], several authors suggest a series of consecutive morphological operations [Soler 2000, Lin 2006]. Other approaches rely on object recognition [Brejl 2000] or a priori knowledge about typical positions of the sought organ in the CT volume [Heimann 2006] or combine a priori knowledge with morphological operations [Tsaagan 2002]. While these approaches work well for specified organs, they cannot be generalized for other segmentation tasks. In order to come up with a generalizable solution, de Brujine and Nielsen proposed an automatic initialization of the template employing shape particle filtering [de Bruijne 2004] for 2D segmentation. A similar approach applied to 3D segmentation based on a global-search in the image was proposed by Heimann et al. [Heimann 2007b]. The algorithm uses the principal ideas of evolutionary programming [Fogel 1966] and evolutionary strategies [Schwefel 1995] in order to determine the optimal placement of the model. The algorithm consists of the following steps:

1. A random set of normally distributed affine transformations T_k and deformations Ω_k is generated with $k = [1, ..., N]$.

2. By applying Ω_k and T_k to the mean shape of the model, a random population of shapes $R = \{S_1, ..., S_N\}$ is built.

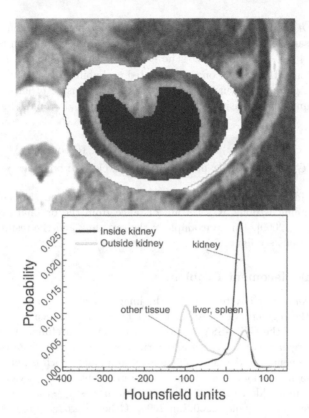

Figure 5.1: *Estimated grey value density functions for the inside (black) and the outside (white) region of the kidney using a Parzen window approach.*

3. The best qualified (or *fittest*) individuals $\hat{R}_k$ of the random population are selected.

4. For each $\hat{R}_k$, the transformation $\hat{T}_k$ as well as the deformation $\hat{\Omega}_k$ are modified randomly and again applied to the mean shape of the model to generate a new (better) population of shapes.

5. This is iterated until a good initial position and a good initial mean shape deformation are found.

The quality of placement is measured by comparing model-specific features to the features in the image. For an example of a random shape population generated for the GGM-SSM of the kidney please refer to figure 5.2.

For our experiments, the means of the normal distributions for the transformation as well as for the deformation equal zero. The standard deviation for $p(T)$ is determined

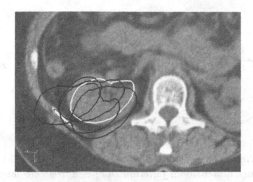

Figure 5.2: *Five examples of a random population of shapes generated for the GGM-SSM of the kidney in a CT image. The white contour belongs to the randomly deformed mean shape which serves as input for the next iteration.*

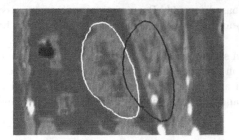 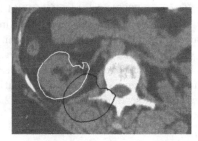

Figure 5.3: *Automatic initial placement. Example of the result of the automatic evolutionary algorithm: original mean shape of the GGM-SSM (black) and final best fit (white).*

heuristically while the standard deviations for $p(\Omega) = \{\omega_1, ..., \omega_n\}$ are the standard deviations $\{\lambda_1, ..., \lambda_n\}$ of the GGM-SSM as computed in section 3.5.2. The model-specific features evaluated in order to measure the fitness depend on probability of points lying on the boundary of the organ. This is measured by the sum of distances between GGM-SSM points and the nearest voxel with high image gradient magnitude which reliably led to good initial placement results. For an example, see figure 5.3.

5.2 The GGM-SSM in Implicit Function Segmentation

In this section, a method is developed for integrating the GGM-SSM into an implicit segmentation scheme. An implicit segmentation scheme has several advantages over an explicit one: First, no remeshing algorithms need to be implemented. Moreover, it is easy to integrate regional statistics as e.g. grey value distribution models and finally,

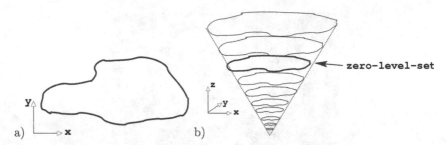

Figure 5.4: *Embedding level set function. a) Contour in 2D. b) The same contour embedded in the higher dimensional function $\phi(x) \in \mathbb{R}^3$ as zero level set at $\phi(x) = 0$.*

they are very flexible topologically. A comprehensive review about the advantages of level set methods in medical image segmentation can be found in the work of Cremers et al. [Cremers 2007]. As the GGM-SSM is based on a MAP estimation and is computed by a global criterion, the integration into an implicit segmentation framework can be realized in a closed mathematical form.

This chapter is organized as follows: In section 5.2.1, the mathematical background of level set methods and their application to implicit segmentation is summarized. The development of the MAP estimation and its solution by an energy functional is presented in section 5.2.2. Sections 5.2.3 and 5.2.4 are dedicated to the derivation and optimization of the energy functional.

5.2.1   Segmentation Using Level Sets

As explained in the section about deformable models (section 2.4.1), the segmentation problem in the variational framework is formulated as the minimization of an energy functional $E(\Gamma)$ with respect to the contour Γ. The key idea is to move the contour in direction of the negative energy gradient $-\frac{\partial E(\Gamma)}{\partial \Gamma}$. In implicit function segmentation, commonly the contour is embedded as the zero level set of a higher dimensional function over the image space $\phi : \Omega \to \mathbb{R}$:

$$\Gamma = \{x \in \Omega | \phi(x) = 0\},$$

see figure 5.4. Most commonly, the front propagation of the contour is realized by evolving the embedding function ϕ using level set methods [Dervieux 1979, Osher 1988, Malladi 1995]. Instead of minimizing the functional defined on the space of contours directly as done e.g. by Caselles et al. [Caselles 1993], several authors propose to embed $E(\Gamma)$ into the variational framework described by $E(\phi)$ in order to search for the level set function $\hat{\phi}$ whose zero level set best describes the organ boundary

[Zhao 1996, Chan 2001]:

$$\hat{\phi}(x) \begin{cases} > 0 & \forall x \text{ outside the organ} \\ = 0 & \forall x \text{ on the boundary} \\ < 0 & \forall x \text{ inside the organ} \end{cases}$$

In that case, $E(\phi)$ can be minimized using the Euler-Lagrangian equation

$$\frac{\partial \phi}{\partial t} = -\frac{\partial E(\phi)}{\partial \phi}$$

where the artificial time $t > 0$ is introduced for parameterizing the descent direction. We solve the derivation by computing the gradient descent

$$\phi^{t+1} = \phi^t - h \frac{\partial E(\phi)}{\partial \phi}$$

with $h > 0$ as the step size.

In the literature of medical image analysis, implicit function segmentation has been applied efficiently e.g. to the detection of a fetus in ultrasound images [Caselles 1997], of the femur in MR images [Leventon 2000a], of the corpus callosum in MR images [Leventon 2000a], of glioma in MR images [Droske 2001], of the left ventricle in cardiac MR images [Tsai 2003], of the prostate of pelvic MR images [Tsai 2003], of lateral brain ventricles in MR images [Rousson 2004] and of the liver in four-dimensional CT images [Schmidt-Richberg 2009].

5.2.2   MAP Estimation on the Level Sets

As shown in the work of Paragios and Deriche [Paragios 2002], the segmentation problem can be formulated in a probabilistic framework where the a posteriori probability $p(\mathcal{P}(X)|I)$ of an optimal partitioning $\mathcal{P}(X)$ given the image I is maximized. Based on this principle, in this thesis a maximum a posteriori estimation is developed of a level set function ϕ whose zero level set best separates the organ from the background under a shape constraint introduced by the GGM-SSM. This leads to a unified statistical framework which is presented in detail in this section.

Given a shape represented as a set of points with model parameters Θ in our GGM-SSM, we first model the probability of a surface with respect to that shape. This amounts to specifying the probability of a function ϕ whose zero level set is the object boundary knowing the GGM-SSM deformation parameters $Q = \{T, \Omega\}$ (The model parameters are detailed in section 3.4). This is the first step. For the next step, we work with the following image formation model: The intensity is assumed to follow a law p_{in} for the voxels inside the object and a law p_{out} for the voxels outside the object. Given this generative model, the segmentation is the inverse problem: The MAP method consists of estimating the most probable parameters ϕ and Q given the observation of an image $I : X \to \mathbb{R}$. Hence, the level set function ϕ is evolved such that $p(\phi, Q|I)$ is maximized:

$$MAP = \operatorname{argmax} p(\phi, Q|I) \quad = \quad \operatorname{argmax} \frac{p(I|\phi, Q)p(\phi|Q)p(Q)}{p(I)}.$$

The shape prior does not add any information when the zero level set of ϕ is known, so I and Q are conditionally independent events $p(I|Q, \phi) = p(I|\phi)$, and we can write

$$p(\phi, Q|I) \quad = \quad p(\phi, T, \Omega|I) = \frac{p(I|\phi)p(\phi|T, \Omega)p(T, \Omega)}{p(I)}.$$

The probability $p(I)$ is constant for a given image. Besides, the probability of the transformation $p(T)$ is assumed to be independent and uniform, so we derive the following energy functional:

$$E(\phi, Q) \quad = \quad -\alpha \log(p(I|\phi)) - \tau \log(p(\phi|Q)) - \kappa \log(p(\Omega)) \tag{5.1}$$

with introduced weights $\alpha, \kappa, \tau \in \mathbb{R}$ to normalize the scale of the distributions. The first term of equation (5.1) describes the region-based energy with object specific priors which are given by the normalized grey value distributions p_{in} inside the organ and p_{out} outside the organ as found in the training data set which leads to

$$\log(p(I|\phi)) \quad = \quad -\int_X (1 - H_\epsilon(\phi(x))) \log p_{in}(I(x))dx - \int_X H_\epsilon(\phi(x)) \log p_{out}(I(x))dx.$$

The function $H_\epsilon(\phi(x))$ is a continuous approximation of the Heaviside function which is close to one outside the object and close to zero inside the object. The regularization of H are chosen as proposed in [Zhao 1996]:

$$H_\epsilon(\phi) = \begin{cases} 1 & \text{if } \phi(x) > \epsilon \\ 0 & \text{if } \phi(x) < -\epsilon \\ \frac{1}{2}\left[1 + \frac{\phi(x)}{\epsilon} + \frac{1}{\pi}\sin(\frac{\pi\phi(x)}{\epsilon})\right] & \text{if } |\phi(x)| \leq \epsilon \end{cases} \tag{5.2}$$

For an illustration of the approximated curve see figure 5.5.

The second term represents the front propagation of ϕ guided by the GGM-SSM which models all points x as a mixture of Gaussian measurements of the (transformed) model points m_j. Following our EM-ICP principle introduced in section 3.2, the probability of a point x modeled by the GGM-SSM given Q is the normalized sum of correspondence probabilities of x and all m_j and equals

$$p(x|Q) = p_\Theta = \frac{1}{N_m} \sum_{j=1}^{N_m} \exp(-\frac{|x - T \star m_j|^2}{2\sigma_\Theta^2}).$$

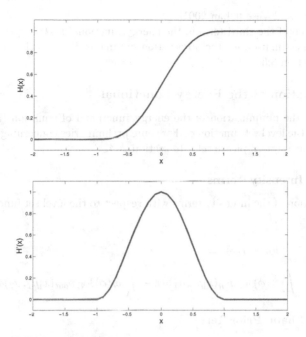

Figure 5.5: *Regularization of the Heaviside function (top) using equation (5.2) and the associated delta function δ_ϵ with support $\epsilon = 1$.*

In the following, p_Θ denotes the probability given by a GGM-SSM with model parameters $\Theta = \{\bar{M}, v_p, \lambda_p, n\}$ which means that Θ is fixed. The probability of a point x with respect to the model described by Θ then depends on the observation parameters $Q = \{T, \Omega\}$. The parameters are used as defined in section 3.3.1.

For a contour Γ describing the zero level set of ϕ, the log of the probability is computed by $\log(p(\phi|Q)) = \log(\prod_{x \in \Gamma} p(x|Q)) = \int_{x \in \Gamma} \log p(x|Q)dx$. The integration over the whole length of the contour is then expressed by

$$log(p(\phi|Q)) = \int_X \delta_\epsilon(\phi(x))|\nabla \phi(x)| \log p_\Theta dx, \qquad (5.3)$$

with $\delta_\epsilon(\phi(x))$ having a small support > 0. Then a normalization is added over the length which leads to $\log(p'(\phi|Q)) = \log(p(\phi|Q)p(\phi|l_0)) = \int_X \delta_\epsilon(\phi(x))|\nabla \phi(x)|$ $(\log p_\Theta - \beta)dx$ with $\beta = \frac{1}{l_0} \in \mathbb{R}$ where l_0 controls the normalization of the length. For $p_\Theta = const$ this equation is generalized to the classical smoothing term

$$\int_X \delta_\epsilon(\phi(x))|\nabla \phi(x)|dx$$

as used by Chan and Vese [Chan 2001].

The definition of the third term in the energy functional $p(\Omega)$ is given by the maximum likelihood estimation for the observation parameter Ω given the model, see equation (3.8) in section 3.3.1.

5.2.3 Derivation of the Energy Functional

In this section, the minimization of the energy functional of equation (5.1) is derived with respect to the level set function ϕ. For some preliminaries concerning mathematical rules used in this section, please refer to section A.4.

5.2.3.1 The Intensity Terms

The differentiation of the intensity terms with respect to the level set function ϕ is quite easy as $\frac{\partial}{\partial \phi} H_\epsilon(\phi) = \delta_\epsilon(\phi)$:

$$\frac{\partial}{\partial \phi} \log(p(I|\phi)) =$$
$$\int_X \delta_\epsilon(\phi) \log p_{in}(x|\mu_1, \sigma_1) dx - \int_X \delta_\epsilon(\phi) \log p_{out}(x|\mu_2, \sigma_2) dx \qquad (5.4)$$

5.2.3.2 The Shape Prior Term

The differentiation of the shape prior term $E_\Theta(\phi) = log(p(\phi|Q))$ as formulated in equation (5.3) with respect to ϕ is a bit tricky. For one thing, we have to deal with the derivative of the Dirac distribution δ'_ϵ. The solution is based on the principle of directional derivatives and integration by parts. The aim is to determine the differential coefficient of $E_\Theta(\phi)$, so we first introduce the function $\alpha : X \to \mathbb{R}$. In order to compute

$$E_\Theta(\phi + \eta\alpha) = \int_X \log p_\Theta \delta_\epsilon(\phi + \eta\alpha) |\nabla\phi + \eta\nabla\alpha| dx.$$

with $\eta \to 0$, we use the Taylor development for a linearization of the delta distribution $\delta_{epsilon}$ at point $(\phi + \eta\alpha)$ and write

$$E_\Theta(\phi + \eta\alpha) = \int_X \log p_\Theta \left(\delta_\epsilon(\phi) + \eta\delta'_\epsilon(\phi)\alpha\right) |\nabla\phi + \eta\nabla\alpha| dx.$$

Using the equation $|\nabla\phi + \eta\nabla\alpha| = |\nabla\phi| + \eta\frac{\nabla\phi^T \nabla\alpha}{|\nabla\phi|} + O(\eta^2)$ which is derived from the binomial series in equation (A.7) allows to write $E_\Theta(\phi + \eta\alpha)$ as a sum of $E_\Theta(\phi)$ and additional terms:

$$E_\Theta(\phi + \eta\alpha) = \int_X \log p_\Theta \left(\delta_\epsilon(\phi) + \eta\delta'_\epsilon(\phi)\alpha\right) \left(|\nabla\phi| + \eta\frac{\nabla\phi^T \nabla\alpha}{|\nabla\phi|} + O(\eta^2)\right) \qquad (5.5)$$
$$= E_\Theta(\phi) + \eta\int_X \log p_\Theta \, \delta'_\epsilon \cdot \alpha|\nabla\phi| + \eta\int_X \log p_\Theta \, \delta_\epsilon(\phi)\frac{\nabla\phi^T \nabla\alpha}{|\nabla\phi|} + O(\eta^2).$$

We reformulate the last term of this equation using the product rule of the divergence as stated in equations (A.5) and (A.6). We set $\nabla g = \nabla \alpha$ and $V = \log p_\Theta \, \delta_\epsilon(\phi) \frac{\nabla \phi}{|\nabla \phi|}$. Assuming that there are no objects outside the image, after several derivations we obtain

$$\int_X <\nabla g, V> = -\int_X g \cdot div(V)$$

which is

$$\int_X \delta_\epsilon(\phi) \log p_\Theta \frac{\nabla \phi^T \nabla \alpha}{|\nabla \phi|} = -\int_X \alpha \cdot div(\delta_\epsilon(\phi) \log p_\Theta \frac{\nabla \phi}{|\nabla \phi|}).$$

With this information, we can rewrite equation (5.5) and obtain

$$E_\Theta(\phi + \eta \alpha) = E_\Theta(\phi) + \eta \int_X \log p_\Theta \, \delta'_\epsilon \cdot \alpha |\nabla \phi| - \eta \int_X \alpha \cdot div\left(\delta_\epsilon(\phi) \log p_\Theta \frac{\nabla \psi}{|\nabla \phi|}\right).$$

(5.6)

We solve the last term by again using the product rule for the divergence stated in equation (A.5). This time we set $g = \delta_\epsilon(\phi)$ and $V = \log p \frac{\nabla \phi}{|\nabla \phi|}$. This leads to

$$\int_X div\left(\delta_\epsilon(\phi) \log p_\Theta \frac{\nabla \phi}{|\nabla \phi|}\right) =$$
$$\int_X \delta_\epsilon(\phi) \cdot div\left(\log p_\Theta \frac{\nabla \phi}{|\nabla \phi|}\right) + \int_X <\nabla(\delta_\epsilon(\phi)), \log p_\Theta \frac{\nabla \phi}{|\nabla \phi|}>.$$

The gradient of $\delta_\epsilon(\phi)$ is computed following equation (A.6):

$$\nabla \delta_\epsilon(\phi) = \begin{pmatrix} \frac{\partial \delta_\epsilon(\phi)}{\partial x} \\ \frac{\partial \delta_\epsilon(\phi)}{\partial y} \\ \frac{\partial \delta_\epsilon(\phi)}{\partial z} \end{pmatrix} = \begin{pmatrix} \delta'_\epsilon(\phi) \frac{\partial \phi}{\partial x} \\ \delta'_\epsilon(\phi) \frac{\partial \phi}{\partial y} \\ \delta'_\epsilon(\phi) \frac{\partial \phi}{\partial z} \end{pmatrix} = \delta'_\epsilon(\phi) \nabla \phi,$$

By inserting this into equation (5.6), we get rid of the $\delta'_\epsilon(\phi)$ terms, so the equation simplifies to

$$E_\Theta(\phi + \nu \eta) = E_\Theta(\phi) - \eta \int_X \alpha \delta_\epsilon(\phi) \cdot div\left(\log p_\Theta \frac{\nabla \phi}{|\nabla \phi|}\right).$$

In order to compute the gradient of E_Θ, we now employ the product rule of equation (A.4), setting $g = \log p$ and $V = \frac{\nabla \phi}{|\nabla \phi|}$, which finally leads to

$$\nabla E_\Theta(\phi) = -\delta_\epsilon(\phi) \cdot div\left(\log p_\Theta \frac{\nabla \phi}{|\nabla \phi|}\right)$$
$$= -\delta_\epsilon(\phi) \log p_\Theta \, div\left(\frac{\nabla \phi}{|\nabla \phi|}\right) - \delta_\epsilon(\phi) <\nabla(\log p_\Theta), \frac{\nabla \phi}{|\nabla \phi|}>. \quad (5.7)$$

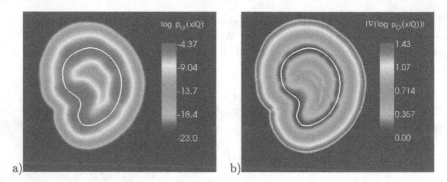

Figure 5.6: *Illustration of the GGM-SSM constraint on the segmentation contour. The GGM-SSM is represented by a white contour slice. a) Log-probability of correspondence for image points x in space. b) Gradient magnitude of log-probability for image points x.*

The constraints of the GGM-SSM on the level set propagation are twofold. The scalar product $< \nabla(\log p_\Theta), \frac{\nabla \phi}{|\nabla \phi|} >$ ensures that the zero level set is actively drawn towards the SSM shape. The values of $\nabla(\log p_\Theta) = \nabla(\log p(x|Q))$ obviously depend on the distance of points x to the GGM-SSM shape. A 2D example is illustrated in figure 5.6(b). The curvature term $\log p_\Theta \; div\left(\frac{\nabla \phi}{|\nabla \phi|}\right)$ ensures that the smoothness factor has more influence on the zero level set evolution at locations of low GGM-SSM probability than at locations with high GGM-SSM probability. This is illustrated in figure 5.6(a). Hence, we use a prior whose contour is length minimizing. The variance σ_Θ^2 of the probability distribution p_Θ is a sensitive parameter and has to be carefully adapted to the problem at hand.

5.2.4 Optimization of the Energy Functional

The derivatives of the energy functional terms derived in the last section are summed up and written in the gradient descent function as

$$\frac{\partial \phi}{\partial t} = \delta_\epsilon(\phi)\left(-\alpha_1 \log(p_{in}) + \alpha_2 \log(p_{out}) - \tau < \nabla(\log p_\Theta), \frac{\nabla \phi}{|\nabla \phi|} > \right.$$
$$\left. + div\left(\frac{\nabla \phi}{|\nabla \phi|}\right)(\beta - \tau \log p_\Theta)\right). \tag{5.8}$$

The minimization of the energy functional in equation (5.1) is then done by alternating the gradient decent for the embedding function ϕ with an update of the parameters T and Ω. The update serves to fit the GGM-SSM to the current zero level set.

The gradient descent is solved by a time-step procedure. In each step, the term $< \nabla(\log p_\Theta), \frac{\nabla\phi}{|\nabla\phi|} >$ has to be updated, thus we need to compute $\nabla(\log p_\Theta) = \frac{\partial}{\partial x} \log\left(\sum_j exp(-\frac{|x-T\star m_j|^2}{2\sigma^2})\right)$. This is simply done by repetitively employing the chain rule which leads to the following explicit GGM-SSM term:

$$< \nabla(\log p_\Theta), \frac{\nabla\phi}{|\nabla\phi|} >=$$

$$\left(\frac{1}{\left(\sum_j exp(-\frac{|x-T\star m_j|^2}{2\sigma^2})\right)} \sum_j \left[exp(-\frac{|x-T\star m_j|^2}{2\sigma^2})\frac{T\star m_j - x}{\sigma^2}\right]\right)^T \frac{\nabla\phi}{|\nabla\phi|}.$$

In order to fit the GGM-SSM to the current zero level set, the optimal transformation T and the optimal deformation coefficients Ω have to be found. The transformation T is computed by

$$\frac{\partial E(\phi, T, \Omega)}{\partial T} = \frac{\partial}{\partial T} \int_X \delta_\epsilon(\phi(x))|\nabla\phi(x)| \log\left(\frac{1}{N_m} \sum_{j=1}^{N_m} \exp(-\frac{|x-T\star m_j|^2}{\sigma_\Theta^2})\right) dx = 0$$

with fixed ϕ and Ω. It suggests itself to make use of the global criterion developed for the GGM-SSM computation in section 3.3.2, equation (3.13). The number of observations is set to one with $k = 1$, and the only observation S_1 is represented by the zero level set of the current ϕ. The affine EM-ICP registration is employed to register the SSM to the zero level set: First the correspondence probabilities between the zero level set and the points of the SSM are established in the expectation step and then T computed in the maximization step as explained in section 3.4.1. Here, the zero level set is represented by all voxels of the level set function where it holds $\delta_\epsilon \neq 0$. The implementation is done efficiently employing sparse fields.

Subsequently, the level set function ϕ and the transformation T are fixed and the deformation coefficients Ω are computed which solve $\frac{\partial E(\phi, \Omega, T)}{\partial \Omega} = 0$. This leads to a matrix formulation in a closed form solution as explained in section 3.4.2 and shown in equation (3.17).

In summary, our implicit segmentation algorithm using the GGM-SSM is implemented as shown in pseudocode 5.1

5.3 Evaluation on Kidney CT Images

In an experimental evaluation, the level set segmentation framework is applied to the segmentation of the left kidney in noisy CT images impaired by breathing artefacts.

Algorithm 5.1 *Pseudocode of implicit segmentation using the GGM-SSM prior*

Place GGM-SSM automatically in image (employing the evolutionary algorithm introduced in section 5.1.2);
Generate initial ϕ based on GGM-SSM;
for $t = 0$ to MAXITER **do**
    Compute $\tilde{\phi}$ according to equation (5.8);
    Update level set: $\phi^{t+1} \leftarrow \phi^t + \tilde{\phi}$;
    Compute GGM-SSM parameters T, Ω (optimizing equation (3.13) with $k = 1$ and S_1 represented by the zero level set of ϕ^{t+1});
    Update GGM-SSM: $M^{t+1} = T \star (\bar{M} + \sum_p \omega_p v_p)$;
end for

The kidneys are a typical organ at risk for cancer radiotherapy in the upper abdomen. They are exposed to irradiation during the treatment of malignant tumor types like carcinoma of the cervix or carcinoma of the pancreas. Thus, an exact segmentation of the kidney helps to reduce the possible harm to a minimum. Fully automatic kidney segmentation is not an easy task as the grey value intensity differences between the kidney and neighbouring organs as the liver and spleen are very small. Moreover, the grey value intensities inside the individual kidney volumes are not very homogeneous which is partly due to the big kidney vessels which are darker than the organ itself and partly due to the poor quality of the abdominal CT images. For an example of the kidney images see figure 5.7.

Most algorithms for (semi-)automatic kidney segmentation from mostly low resolution CT images consist of two steps: First, for automatic initialization, a region in the image is selected where the probability of kidney tissue appearance is high. Second, a local search algorithm is employed in order to detect the kidney contour. Recently published methods using deformable models include the combination of grey level appearance of the target with statistical information about the shape [Tsaagan 2002] or the training of a non-parametric histogram estimate specifying the kidney location [Broadhurst 2006]. Another method proposes the concatenation of different image processing operations as region growing and landmark determinations [Lin 2006]. Looking at the evaluations, all of those methods lead to volume overlaps around 0.88 (where it is not clear which measuring coefficients were used) and an average surface distance of $1mm$ [Broadhurst 2006] and respectively around 1 voxel with resolution $0.63 \times 0.63 \times 10mm^3$ [Tsaagan 2002] between the results and the gold standard. All papers report failure of their method for some cases which were mainly accounted for to poor quality of the automatic location initialization.

5.3.1   Segmentation Experiment

Kidney GGM-SSM: Our training data set consists of 16 CT images of the abdominal region which were taken from healthy live liver donors. The data set as well as the

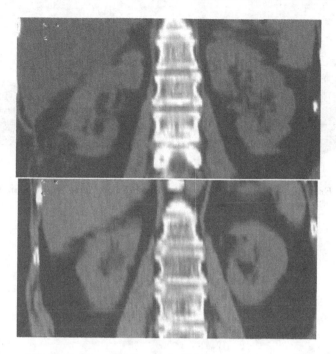

Figure 5.7: *Examples of abdominal CT images including the kidney.*

associated segmentations of the left kidney were kindly provided by the Department of Computer Science, UNC, Chapel Hill. The segmentations were performed by medical students. The size of the images is $512 \times 512 \times (32 - 52)$ voxels with resolution $0.98 \times 0.98 \times (2.9 - 5.0)mm^3$ where the kidney measures about $75 \times 60 \times 100mm^3$. The GGM-SSM for the kidney is built using a training data set of 10 segmented observations. For some observation examples see figure 5.8. The segmentation method is then tested on the remaining 6 kidneys. For computing the GGM-SSM, the global criterion (equation (3.13)) is optimized as elaborated in section 3. The algorithm multi-scale parameters (described in section 3.6) are set to $\sigma = 20mm$, reduction factor $= 0.9$, number of iterations $= 20$. The resulting kidney GGM-SSM can be seen in figure 5.9 where the mean shape and the deformations according to the first and second modes of variation are depicted.

Distribution Model: For our application on the estimation of p_{in} and p_{out}, the Parzen window approach described in section 5.1.1 is employed. The intensities around the kidney surfaces of our training data set which are coded by the Hounsfield scale are sampled. A Gaussian kernel and a width of $h = 5$ are used, see figure 5.1.

Figure 5.8: *Examples of surface representations of segmented kidneys in the training data set.*

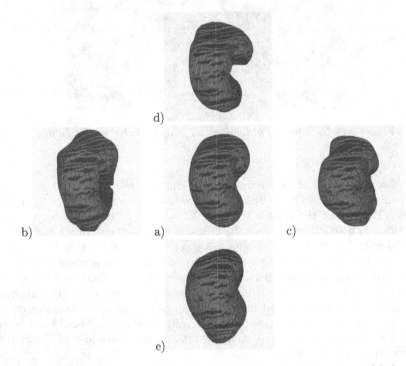

Figure 5.9: *GGM-SSM computed for a training data set of 10 segmented kidneys. (a) shows the mean shape, (b-e) show the mean shape deformed with respect to first and second mode of variation:* $\bar{M} - \lambda_1 v_1$, $\bar{M} + \lambda_1 v_1$, $\bar{M} - \lambda_2 v_2$, $\bar{M} + \lambda_2 v_2$.

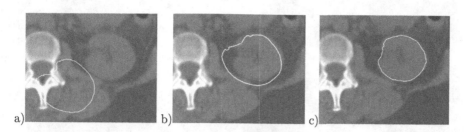

Figure 5.10: *GGM-SSM during segmentation a) The GGM-SSM is placed in the image. b) The GGM-SSM is automatically initialized to its starting position. c) The GGM-SSM deforms under the optimization of the global criterion.*

Set-Up: In order to evaluate the influence of the shape prior term, the results of our algorithm are compared with the results of the segmentation algorithm proposed by Schmidt-Richberg et al. who use a very similar energy functional but without a shape prior term [Schmidt-Richberg 2009]. Each data set is segmented once with the level set segmentation without shape priors as proposed by Schmidt-Richberg et al. and once with the GGM-SSM prior information integrated in the level set segmentation as developed in section 5.2. The algorithm is implemented as shown in pseudocode 5.1. For the segmentation, the weights are set to $\alpha_1 = 1$, $\alpha_2 = 1$, $\kappa = 1$, $\beta = 0$ and $\tau = \{0.1, 0.2\}$. In most cases, the algorithm converged after 150 iterations. For both methods, the same distribution model is used. For an example of the GGM-SSM deformation during the segmentation steps please see figure 5.10.

Results: The results are compared to the gold standard segmentations by evaluating the Jaccard coefficient, the Dice coefficient and the Hausdorff distance, see table 5.1. Both level set frameworks using a-priori information on the grey level intensities yields good segmentation results overall. The SSM constraint on the level set evolution yields even better results in all cases. The advantage of adding the prior shape information can be seen distinctly for patient 2 where the Hausdorff distance diminished from 9.95mm to 5.0mm and for patient 6 where the Hausdorff distance diminished from 12.57mm to 7.68mm. This is due to the fact that the evolving zero level is attracted by neighbouring organs with similar grey value intensities as the kidney. The Hausdorff distance can be seen as an indicator for the leakage risk. This leakage can be successfully prevented by integrating the SSM prior on shape probabilities. As an example, the effect on patient 2 is shown in figure 5.11(b).

5.3.2 The Role of the Parameters

As our energy functional in equation (5.1) is derived by a MAP explanation, in theory all coefficients should be equal to 1. Expanding on this probabilistic analogy, the traditional

		only LS	LS + SSM
	D(A,B)	0.93	0.93
Pat1	J(A,B)	0.88	0.87
	H(A,B)	8.66	6.40
	D(A,B)	0.91	0.93
Pat 2	J(A,B)	0.83	0.88
	H(A,B)	9.94	5.0
	D(A,B)	0.89	0.91
Pat 3	J(A,B)	0.81	0.84
	H(A,B)	5.83	5.10
	D(A,B)	0.88	0.89
Pat 4	J(A,B)	0.78	0.80
	H(A,B)	8.01	6.40
	D(A,B)	0.92	0.92
Pat 5	J(A,B)	0.86	0.86
	H(A,B)	4.58	4.24
	D(A,B)	0.84	0.86
Pat 6	J(A,B)	0.73	0.75
	H(A,B)	12.57	7.68

Table 5.1: *Segmentation Results for six different data sets. Left: Level set segmentation without GGM-SSM shape prior as done with the algorithm of Schmidt-Richberg et al. [Schmidt-Richberg 2009]. Right: Level set segmentation using the GGM-SSM shape prior as developed in section 5.2.2. D(A,B): Dice coefficient. J(A,B): Jaccard coefficient. H(A,B): Hausdorff distance in mm.*

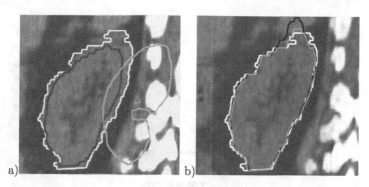

Figure 5.11: *Segmentation results on a kidney in CT data, sagittal slice. The white contour is the gold standard segmentation. Image (a) shows the initial contour in light grey and the contour after applying the automatic evolutionary algorithm as described in section 5.1.2 in black. Image (b) shows the result of the unconstrained (black) and the result of the SSM constrained (light grey) level set segmentation. The black contour leaked into the adjacent organ (liver).*

coefficients of the variational methods (as e.g. in [Chan 2001] or [Rousson 2004]) can be seen as powering factors which flatten or peak the density distributions. Concerning the GGM-SSM term (equation (5.3)), the standard deviation σ_Θ controls the matching of the GGM-SSM to the zero level set. This means that in practice, σ_Θ should have values around $5mm$ to guarantee a successful matching for the problem at hand as this is the mean point distance in the model. However, the value of σ_Θ also controls the strictness of the spatial constraint, so the introduction of the coefficients τ, β and α is necessary in order to position the influence of the SSM with respect to the other terms. What is more, β can be equal to 0 because the smoothness term $div\left(\frac{\nabla\phi}{|\nabla\phi|}\right)$ is also governed by τ as can be seen in equation (5.8). Moreover, employing $-\tau \log p_\Theta$ as weight has the advantage of using a distance-dependent smoothing term. Figure 5.12(a) shows the influence of the choice of σ_Θ for the Hausdorff distances obtained in the segmentation experiments with $\alpha = 1, \beta = 0$ and $\tau = 0.1$. These parameters lead to satisfying results for all kidneys except kidney 1. The optimal values for σ_Θ are similar for all kidneys and should not exceed $5mm$ in this case.

The relation between the parameters τ and σ_Θ are illustrated in figure 5.12(b) where the Hausdorff distances for two kidney segmentations are plotted with respect to σ_Θ for different values of τ. For a smaller τ the optimal σ_Θ becomes smaller as well which results in a left shift of the curve. This is due to the fact that a smaller σ_Θ as well as a greater τ result in a stricter constraint of the level set front propagation. However, the best result for the Hausdorff distance remains the same for both choices of τ.

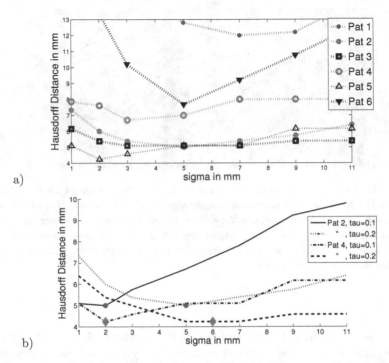

Figure 5.12: *Hausdorff distances. a) shows the Hausdorff distances of the segmentation results under parameters $\alpha = 1$, $\beta = 0$ and $\tau = 0.1$ for all kidneys with respect to σ_Θ. b) illustrates the relation between the parameters τ and σ_Θ and their influence on the resulting Hausdorff distances.*

5.4   Multiple Shape Class Segmentation

On the grounds that shape, size and location of neighbouring anatomical structures influence each other directly and indirectly, a thriving strategy is the extension of the region of interest for the segmentation to adjacent structures. The integration of these geometric relation information about adjoining structures as a priori knowledge renders a segmentation algorithm a lot more robust. This idea can be exploited for example in an attempt to simplify segmentation processes for low-contrasted structures as shown e.g. by Palm et al. who use a balloon model coupled to a SSM to find the vocal cord and utilize the results to find the glottis next [Palm 2001]. Costa et al. present a coupled segmentation framework employing an explicitly represented SSM of the prostate for segmenting the bladder and prostate simultaneously [Costa 2007]. In [Zeng 1999], the segmentation of the cortex from 3D MR images is performed by a coupled surface propagation. This is realized by coupling the segmentation results of two adjacent borders of the cortex by verifying that the distance between the borders does not exceed a certain interval. Pitiot et al. enhance this idea by constructing deformable models for different brain structures and regulating the associated segmentations by a distance map which determines certain distance values that have to hold between the structures [Pitiot 2005]. In another approach, Ciofolo et al. model the distances between brain structure contours as a fuzzy variable so to avoid overlapping between contours of different level sets [Ciofolo 2005]. A very interesting method is proposed by Tsai et al. who employ multiple signed distance functions as implicit representations of multiple shape classes within the image [Tsai 2004]. By doing a PCA on these functions they then obtain a coupling between the multiple shapes within the image and hence effectively capture the co-variations among the neighbouring structures. Implicit function segmentation is topologically flexible and therefore well suited to segment non-spherical topologies as well as objects containing multiple shape classes. As our GGM-SSM prior is able to model non-spherical anatomies and also anatomies consisting of more than one structure, our aim is to extend the segmentation algorithm presented in section 5.2 for such kind of segmentation. Section 5.4.1 is dedicated to the mathematical adaption of the GGM-SSM to multiple object modeling and its integration into the time step procedure of the segmentation scheme. In section 5.4.2, first experiments are done on acetabulum and femoral head data which feature a non-spheric anatomy and consist of two non-connected structures.

5.4.1 Development of the Algorithm

5.4.1.1 Extension of the GGM-SSM to Multiple Structures

For the segmentation of more than one shape class, the shape prior has to represent a training data set of multiple-structure observations. In order to model multiple structures using only one GGM-SSM, an overlap between structures belonging to different shape classes has to be avoided. Therefore, the EM-ICP registration used for aligning the model with the observations has to be adapted to that task. To recap: for one structure, the correspondence probability between an observation point s_{ki} and a model point m_j reads:

$$\gamma_{ijk} = \frac{\exp\left(-\frac{\|s_{ki}-T_k \star m_{kj}\|^2}{2\sigma^2}\right)}{\sum_{l=1}^{N_m} \exp\left(-\frac{\|s_{ki}-T_k \star m_{kl}\|^2}{2\sigma^2}\right)}$$

as explained in section 3.3.2. On the one hand, the objective is to compute one transformation which transforms two or more structures together in order to keep their spatial relationship. On the other hand, an overlap of structures of different types has to be avoided to guaranty a good modeling. To do so, it has to be made sure that the correspondence probability $\gamma_{ijk} = 0$ if points m_j and s_{ki} belong to different structures. This

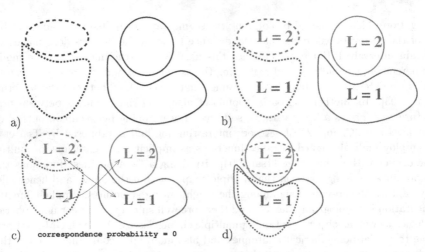

Figure 5.13: *EM-ICP for multiple structure observations. a) Observations consisting of two structures. b) Structures are labeled $L = 1$ and $L = 2$. c) Points belonging to structures with different labels have a correspondence probability of zero. d) Aligned observations.*

is done by labeling the points congruently over the whole training data set and then computing

$$\gamma_{ijk} = \begin{cases} 0 & if\ L(m_j) \neq L(s_{ki}) \\ \dfrac{\exp\left(-\dfrac{\|s_{ki}-T_k \star m_{kj}\|^2}{2\sigma^2}\right)}{\sum_{l=1}^{N_m} \exp\left(-\dfrac{\|s_{ki}-T_k \star m_{kl}\|^2}{2\sigma^2}\right)} & else \end{cases} \tag{5.9}$$

with $L = \{1, 2, ...\}$ being the label of the respective structures. For an illustration see figure 5.13. Using the labeled correspondence matrix in the EM-ICP registration has the effect that only point pairs belonging to the same shape class guide the registration. The resulting transformation then tries to align the respective structures without causing an overlap inside the observation.

5.4.1.2   Extension of the Segmentation Method to Multiple Structures

The goal is to extend the segmentation algorithm described in section 5.2 (equation (5.1)) for multiple-structure observations. As explained above, only one GGM-SSM is used to model the multiple-structure shape. However, a separate level set function ϕ_L is defined for each structure. This is done for two reasons: First, it allows us to define grey value probabilities p_{in}^L and p_{out}^L for each structure. Secondly, additional anatomical constraints can be defined as for example in case of different shape structures lying close to each other, it is of great interest to prevent separate structures from merging.

The evolution of each level set function is computed by a separate gradient descent using the formulation of equation (5.8). Here, the shape priors in each gradient descent are represented by the respective structures of the GGM-SSM. Importantly, the update of the GGM-SSM is done with respect to all zero level sets with $\phi_{all} = min\{\phi_1, \phi_2, ...\}$, and this step therefore links the evolution of the separate level sets.

The implementation of the multiple-structure segmentation is presented in pseudocode 5.2.

Algorithm 5.2 *Pseudocode of implicit two shape class segmentation using the GGM-SSM prior*

Place GGM-SSM automatically in image (employing the evolutionary algorithm introduced in section 5.1.2);
Generate initial ϕ_1 and ϕ_2 based on GGM-SSM;
Set d as minimal allowed distance between the two level sets;
for $t = 0$ to MAXITER **do**

Compute $\tilde{\phi}_1$ according to equation (5.11);
Update level set;
{Apply constraint:}

$$\phi_1^{t+1} = \begin{cases} \phi_1^t + 0 & \text{if } \phi_2^t(x) < d \\ \phi_1^t + \tilde{\phi}_1 & \text{else} \end{cases} ;$$

Compute $\tilde{\phi}_2$ according to equation (5.11);
Update level set;
{Apply constraint:}

$$\phi_2^{t+1} = \begin{cases} \phi_2^t + 0 & \text{if } \phi_1^{t+1}(x) < d \\ \phi_2^t + \tilde{\phi}_2 & \text{else} \end{cases} ;$$

Form one contour: $\phi^{t+1} = min\{\phi_1^{t+1}, \phi_2^{t+1}\}$;
Compute GGM-SSM parameters T, Ω (optimizing equation (3.13) with $k = 1$ and S_1 represented by the zero level set of ϕ^{t+1});
Update GGM-SSM: $M^{t+1} = T \star (\bar{M} + \sum_p \omega_p v_p)$;
end for

The Boundary Term:

For organs whose grey value intensity differs significantly from the background's as is the case e.g. for bones, the gradient information in the image could be interesting to be exploited for the segmentation. To do so, an edge term is added to the energy functional described in equation (5.1) which serves to actively draw the zero level set towards organ boundaries. Based on the Geodesic Active Region model proposed by Paragios and Deriche [Paragios 2002], an energy functional based on the boundary term can be introduced by

$$E_{boundary}(\phi) = \int_X \delta_\epsilon g(I) |\nabla \phi| dX$$

where

$$g(I) = \frac{1}{1 + |\nabla(G_\sigma * I)|}$$

with G_σ being a monotonically decreasing function (in our case a Gaussian function). The derivative of the boundary term with respect to the level set function ϕ is complex. It is computed analogously to the derivative of the shape prior as elaborated in section 5.2.3. This finally results in

$$\nabla E_{boundary}(\phi) = -\delta_\epsilon(\phi)g \; div \left(\frac{\nabla \phi}{|\nabla \phi|} \right) - \delta_\epsilon(\phi) < \nabla g, \frac{\nabla \phi}{|\nabla \phi|} > . \qquad (5.10)$$

This term is integrated into the gradient descent of equation (5.8) which leads to the extended gradient descent

$$\frac{\partial \phi}{\partial t} = \delta_\epsilon(\phi) \left(-\alpha_1 \log(p_{in}) + \alpha_2 \log(p_{out}) - \tau < \nabla(\log p_\Theta), \frac{\nabla \phi}{|\nabla \phi|} > \right.$$
$$\left. -\eta < \nabla g, \frac{\nabla \phi}{|\nabla \phi|} > + div \left(\frac{\nabla \phi}{|\nabla \phi|} \right) (\beta - \tau \log p_\Theta - \eta g \,) \right). \qquad (5.11)$$

with $\eta \in \mathbb{R}$ as the associated weight.

The integration of the boundary term is also advantageous when segmentating two or more neighbouring structures simultaneously as the leakage risk might be reduced.

5.4.2   Experimental Evaluation on Hip Joint CTs

A first experimental evaluation is done on hip articulation data. These are well suited for our needs as they feature two shape classes (acetabulum and femoral head) as well as a non-spherical topology since the ischium and the pubis bone form a ring. The intensity within the bones is not constant as the interior consists of trabecular bone whereas the outer shell is a compact cortical bone. This intensity variation is a drawback for thresholding techniques. Moreover, the edges might be blurred by artifacts which deteriorates the accuracy of region growing methods. Besides, a considerable amount of noise or blurring often adds to the complications. Especially the tiny space between the femoral head and the acetabulum poses a problem because automatic segmentation methods have difficulties to recognize the adjoining edges as two different units [Westin 1998].

The CT data set used in this experiment consists of 11 images of the hip joint with resolutions around $0.71 \times 0.71 \times 4mm$ and size $512 \times 512 \times (57 - 78)$ voxels. The resolution in z-direction is not high enough to allow a reliable manual detection of the gap between femoral head and acetabulum in many of the images. Therefore, the medical experts who segmented the training data set chose to augment the resolution in z-direction for a better estimation of the gap. These sampled images then feature resolutions around $1 \times 1 \times 1mm$ and size $256 \times 256 \times (228 - 312)$ voxels, see examples in

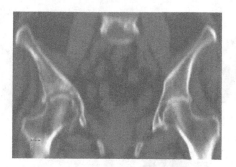 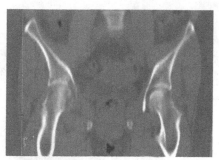

Figure 5.14: *Hip joint CTs: These images belong to the observations which form the training data set.*

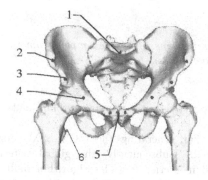

Figure 5.15: *Frontal view of the hipbone and anatomical landmarks. 1-Promontorium, 2-Spina iliaca anterior superior, 3-Spina iliaca anterior inferior, 4-Eminentia iliopubico, 5-Symphyse, 6-Trochanter minor.*

figure 5.14. For each data set one manual segmentation was done by a medical expert. For the evaluation, we are interested in modeling the region of the hip articulation as well as the region with the non-spherical topology. Therefore, the observations are clipped to the region of interest. In order to do a congruent clipping over all observations, the anatomical landmarks on the bones are used as reference (see figure 5.15): The femur is clipped by a horizontal plane cutting $1mm$ below the trochanter minor. The hip bone is clipped by a horizontal plane cutting $5mm$ above the spina iliaca anterior inferior. The results for some of the observations are depicted in figure 5.16. The observations are represented by around 7000 points (minimum 6544 points, maximum 7408 points). In a preprocessing step, a labeling of all observations to distinguish hip bone and femoral head is done where the femoral head is labeled with $L = 1$ and the acetabulum is labeled with $L = 2$. The GGM-SSM for the hip articulation is built using a training data set of 8 observations and the segmentation

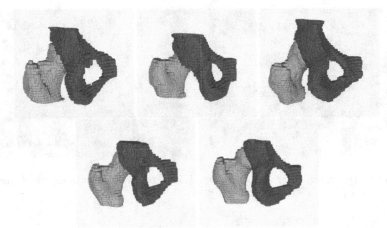

Figure 5.16: *Hip joint observations. These examples from the training data set are labeled to separate femur and hip bone structure.*

method is then exemplarily tested on the remaining 3 hip joints.

Hip joint GGM-SSM: For generating the GGM-SSM, first the barycentres of all observations are aligned. Subsequently, the global criterion (equation (3.13)) is optimized as elaborated in section 3. The algorithm multi-scale parameters (as introduced in section 3.6) are set to $\sigma = 10mm$, reduction factor = 0.9, number of iterations = 15. The resulting hip joint GGM-SSM can be seen in figure 5.17 where the mean shape and the deformations according to the first and second modes of variation are depicted.

Distribution Model: For our application on the estimation of p_{in} and p_{out}, again the Parzen window approach described in section 5.1.1 is used. The intensities are sampled around the bone surfaces of our training data set which are coded by the Hounsfield scale. A Gaussian kernel and a width of $h = 5$ are used, see figure 5.4.2. The intensity distributions for the inside and the outside of the bones greatly overlap especially for the femoral head due to the colour of the bone marrow which resembles the background. This means that the information value of the grey value distribution prior for the segmentation is reduced.

Set-Up: For the segmentation, the weights are set to $\alpha_1 = 0.5$, $\alpha_2 = 0.5$, $\kappa = 1$, $\beta = 0$ and $\tau = \{0.5, 0.8\}$. The cartilage between acetabulum and femoral head measures at its thickest point around $4mm$ (and less in elderly people) and is low-contrasted in the images, so this region is very difficult to segment based on intensity distribution

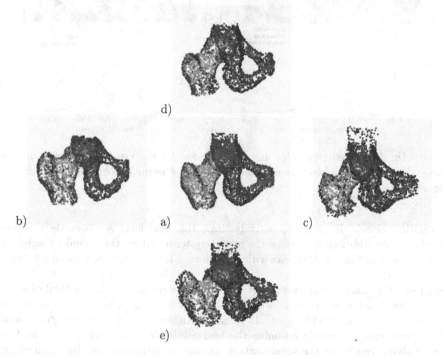

Figure 5.17: *GGM-SSM for the hip joint. a) Mean shape. Deformation along the first (b,c) and second (d,e) variation mode which mainly affect the bulging of the femoral head, the torsion and size of the ischium as well as the CCD angle.*

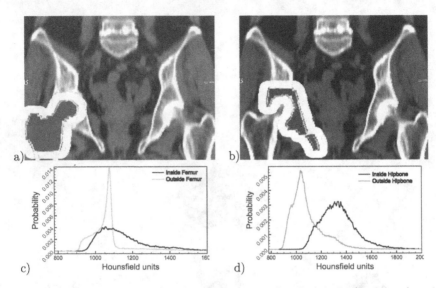

Figure 5.18: *Estimated grey value density functions for the inside (dark grey) and the outside (white) region of the clipped femur (a,c) and hipbone (b,d) using a Parzen window approach.*

information alone. In order to actively draw the zero level set towards the bone boundaries, we additionally employ the boundary term and set the boundary weight to $\eta = 0.3$. The function g is Gaussian with $\sigma = 7mm$. The algorithm is iterated 200 times.

Results: For testing purposes, first we try to segment the hip articulation using the level set segmentation without shape prior employing the algorithm as proposed by Schmidt-Richberg et al. [Schmidt-Richberg 2009]. As the grey values of the bone marrow greatly resemble the background in some regions, this leads to non-satisfying results as the segmentation contour sometimes looses its connectivity. An example for this behaviour is shown in figure 5.19(a) and (b). By integrating the shape prior, these problems could be avoided (see figure 5.19(c)). Two result examples with a close-up on the articulation region are shown in figure 5.20. The shape prior was able to successfully model the non-spherical topology formed by the pubic bone and ischium (see figure 5.21(d)).

Because of the femoral marrow, the zero level set of the implicit function sometimes creates holes inside the femoral structure. Therefore, instead of the Dice coefficient, the surface distance between the deformed GGM-SSM and the expert segmentation is used to asses the evaluation results. These are depicted in table 5.2. The mean distance measures around $3mm$ which seems to be acceptable with regard to the low quality of the data. The distances are illustrated for the hipbone and the femoral

Table 5.2: *Segmentation results. The table shows the mean surface distance and the Hausdorff distance of the final deformed SSM and the manual segmentation in mm.*

	Pat. 1		Pat. 2		Pat. 3	
	Femur	Hipbone	Femur	Hipbone	Femur	Hipbone
mean dist. in *mm*	3.0	2.9	3.5	3.0	2.1	3.1
Hausdorff dist. in *mm*	11.6	12.5	15.8	16.8	16.4	14.3

head in figure 5.21(a) and (b). It becomes clear for patient 2 that the border of the acetabulum posed a problem for the segmentation algorithm. This might be due to the fact that the contrast in that region is very low which is shown in figure 5.21(c). Even for the expert, this region must have been very difficult to detect. In order to validate the results further, inter-individual variability evaluations should be performed in a series with several medical experts.

Overall, the results obtained in this experiment indicate that the method is well suited for two shape class segmentation.

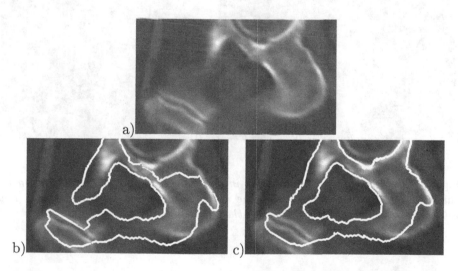

Figure 5.19: *Problematic region for segmentation. Figure a) shows a zoom on the ischium structure of the hip bone where the grey value intensities of bone marrow and background resemble and no clear boundary can be seen. b) Segmentation result of level set segmentation without shape prior. c) Segmentation result of level set segmentation with shape prior.*

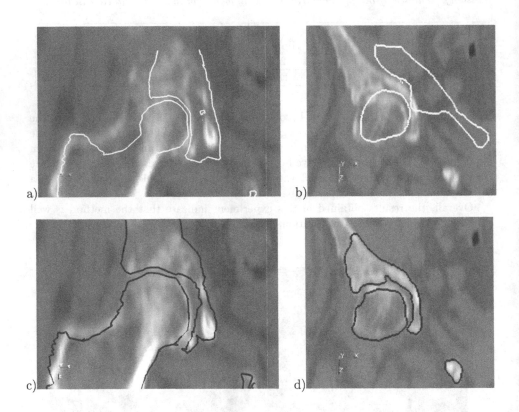

Figure 5.20: *Segmentation results. The images show a view on the segmentation on patient 1 (a,c) and patient 2 (b,d). The initial segmentation is shown in light grey (above) whereas the results are shown in black (below).*

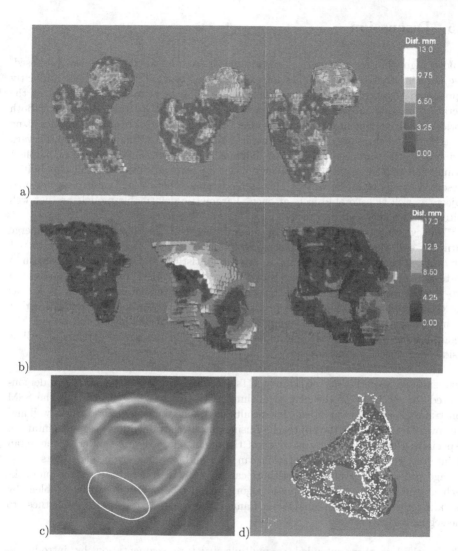

Figure 5.21: *Segmentation Results. a) Surface distances between gold standard and deformed GGM-SSM after segmentation for the hipbones of patient 1, patient 2, patient 3. b) Surface distances between gold standard and deformed GGM-SSM after segmentation for the femoral heads of patient 1, patient 2, patient 3. c) Cut through the acetabulum of patient 2 in CT image. The white ellipse marks the region with low contrast which the segmentation method did not detect well as seen in image (c), middle hipbone. d) Deformed GGM-SSM (white points) during the segmentation of the hipbone.*

5.5   Discussion

A novel algorithm for automatic segmentation of anatomical structures was proposed. The segmentation scheme couples an explicitly represented SSM with an implicitly represented segmentation contour. This approach is new to our knowledge of the literature on this subject and opens new insights on how to take the best of both worlds. Implicit segmentation methods offer several advantages over explicit ones as no remeshing algorithms are needed, the integration of regional statistics is straightforward and finally, they are very flexible to different topologies. Furthermore, an implicit formulation of the segmentation allows to easily take advantage of the capabilities presented by the GGM-SSM: It is able to model non-spherical and multiply-connected objects as well as several objects at once. Parametric deformable models are not well suited for such segmentation tasks.

The evolving contour of implicit models, however, is able to split and merge naturally and allows the simultaneous detection of several objects. In order to put the implicit representation within a unified statistical framework, a maximum a posteriori estimation of a level set was developed. The MAP explanation leads to a two-phase formulation which is optimized based on the image information as well as the GGM-SSM information about probable shapes. This approach is refined further by integrating prior knowledge about grey value distributions inside and outside the organ in order to robustify against intensity inhomogeneities across patients as well as inside the respective structures.

Segmentation experiments on kidney CTs impaired by breathing artefacts demonstrated the efficiency of the new algorithm. Adaptive weights ensure that the SSM constraint is optimally exploited. The results show that the new method works well and improves for some cases the approach of using an unconstrained level set segmentation. Especially when the intensity patterns of the organs close by are similar to the organ of interest, the level set segmentation can leak and produce erroneous results. The leakage problem of level set algorithms can be seen in different segmentation tasks such as the prostate. The proposed algorithm offers a solution to this problem by including the SSMs in a probabilistic framework such that they bring robustness to the segmentation process.

The method is then extended to multiple-structure segmentation by introducing a level set function for each structure. The shape prior information however is modeled by a single GGM-SSM for all structures simultaneously. During segmentation, the evolution of the different level set functions is linked and constrained by the multiple-shape GGM-SSM. Furthermore, by integrating a boundary term into the energy functional, the method is adapted to bone segmentation.

First experiments on hip articulation data indicate that the method is well suited for modeling and segmenting multiple objects at once and also shows that the GGM-SSM is able to be employed as a shape prior for non-spherical anatomies as shown on the example of ischium and pubic bone. Inherently, implicit segmentation techniques are sensitive to the initial placement. This problem gets worse for segmentation of structures lying close-by whose intensities are close. In case of the hipbone articulation segmentation, the grey value distributions from femoral head and hip bone are very similar (see figure 5.4.2). This means that the segmentation will fail if the automatic initial placement positions the initial femoral structure inside the hip structure or vice versa. Therefore, the initial placement has to be controlled carefully.

Even from a low number of samples a prior on the probabilities can be extracted so that no huge training data set is necessary. From a theoretical point of view, a very powerful feature of this method is that a unique criterion is optimized. However, the practical convergence rate has to be investigated more carefully as it depends on the choice of weights in the functional as well as the variance σ_Θ^2 which controls the probability of occurrence with respect to the SSM. In the case of an organ shape which differs greatly from the shapes in the training data set for the SSM, a great sigma is needed in order to not constrain the contour evolution too much (as e.g. for Pat. 1, figure 5.12(a)), so σ_Θ is momentarily used somewhat as interactive parameter which is not the optimal solution. Furthermore, the MAP formulation could be refined by integrating a priori knowledge about the expected volume V_0 which is given by the probability $p(\phi|V_0)$ where V_0 can be determined by evaluating the training data set.

Concerning the method for multiple-structure segmentation, the implementation is currently done using one energy functional for each contour. This approach could be improved by formulating a single energy functional containing all independent level set functions as parameters. The obligatory constraint which forbids an overlap of the independent contours could then be integrated as side condition. Overall, to consolidate the results of multiple-structure segmentation, a more elaborate evaluation on a bigger data set is needed.

CHAPTER 6

Conclusion

Statistical shape models play an important role in medical image analysis, and a wide range of methods well adapted to various applications exists in the literature. The emphasis of this thesis however was not so much to propose a convenient SSM to solve a specific practical problem but to investigate the possibilities of a novel approach to SSM computation. The focus of this manuscript is twofold: First, a novel SSM method was developed in a probabilistic framework. Then, by taking advantage of the particular characteristics of the probabilistic SSM, it was integrated into an implicit segmentation scheme. Both parts were formulated on a sound theoretical foundation and feature new views on well-known problems.

In this chapter, the contributions developed in the course of this manuscript are reviewed and an outlook on possible future research on the subject is given.

6.1 Contributions

6.1.1 Model Computation

As a first step on the path to a novel SSM computation method, an affine extension of the Expectation Maximization - Iterative Closest Point registration algorithm was proposed which directly yields a solution to the fundamental correspondence problem. Here, the observations are represented by unstructured point clouds, and each observation point is modeled as a noised measurement of the model points. This approach actually amounts to representing the surface of the shapes by a mixture of Gaussians. The probabilistic concept offers an intuitive and coherent way to determine correspondences between smooth organ surfaces as well as between shapes where not all observations feature the same prominent shape details. It should be noted that the SoftAssign algorithm [Rangarajan 1997a] offers a related probabilistic formulation but is only justified for a pair-wise registration, not for the group-wise model to observation registration which is required for building the SSM.

The introduction of probabilistic correspondences gives way to a large contribution of this thesis which is the development of a sound mathematical framework for SSM computation presented in chapter 3 and [Hufnagel 2007b, Hufnagel 2008b]. To realize this, the SSM problem has been viewed from the new angle of generative models: Given a set of observations, it has been sought for the model which most probably generated those observations. As the model itself is modeled as a random variable described by a Gaussian distribution, a maximum a posteriori estimation of the whole scene has been formulated. Here, observation and model parameters were unified in one global criterion which has not been done before to the author's knowledge. It could be shown that the optimization of the criterion led to closed form solutions for all parameters except the variation modes which are efficiently solved for iteratively. Since the SSM computation is done by optimizing a global criterion, a theoretical convergence of the algorithm is ensured. Furthermore, in contrast to methods using the principal component analysis, the variation modes of the SSM presented here only model the shape variation and not the noise which is represented separately through the Gaussian Mixture. This implies a possible answer to modeling the uncertainties inherent to surface representations of segmented organs.

Apart from the methodological contributions, the GGM-SSM resulting from the new computation algorithm itself significantly adds to the state-of-the-art. A main advantage is the simplicity of the point-based SSM with respect to its power. The application to an arbitrary training data set is straightforward since no preprocessing to establish correspondences is needed, and the point numbers from observation to observation as well as the point density may vary. As the connectivity between points does not play a role, the GGM-SSM is very flexible to different kinds of topologies and therefore well-suited to model non-spherical or multiply-connected objects as well as several objects at once. The superior quality of the GGM-SSM compared to a classical point-based SSM computed under the use of the iterative closest point algorithm and a principal component analysis (ICP-SSM) could be demonstrated on synthetic and real data sets as presented in chapter 4 and [Hufnagel 2009a]. While the ICP-SSM is a faster method, the GGM-SSM reliably succeeded in capturing shape details as well as extreme shape variations which were lost for the ICP-SSM.

Throughout this thesis, the confidence in surface information for SSM computation is considered arguable as these are only approximations of the true surfaces. Nevertheless, in practice surface-based SSMs obtain useful results. In order to place the new approach in the literature, a comparison of a MDL-SSM and the GGM-SSM was performed on a synthetic data set which proved to be a difficult endeavour as a comparable metric had to be defined. Finally, the results were evaluated using the Jaccard coefficient for which surfaces had to be approximated for the GGM-SSM results. The experiments showed that the GGM-SSM almost reached the performance of the MDL-SSM. The difference is probably due to the fact that in the MDL-SSM points are allowed to freely move over the surfaces so that the results do not depend on the original point distribution in the observation meshes. Unlike the GGM-SSM however, the MDL-approach is constrained to surface representations for spherical topologies.

6.1.2 Segmentation

Another significant contribution of this thesis lies in the development of a novel segmentation algorithm as presented in chapter 5 and [Hufnagel 2009c]. The algorithm integrates an explicitly represented shape prior into an implicit segmentation scheme.

Most implicit segmentation schemes which make use of shape priors do statistics on signed distance maps which do not constitute a linear space. Furthermore, the principal components of implicit shape models describe the variability of the distance maps but not the variation of the embedded contours. Therefore, understanding the variability information on distance functions is not obvious. In contrast, the variability model of a parametric SSM encodes the variation for each point of the model which allows a direct physical interpretation of the shape variability.

The objective in this work was to exploit the advantages offered by implicit segmentation methods without relinquishing the benefits given by explicitly represented SSMs. Since the GGM-SSM was formulated in maximum a posteriori explanation and is computed in a probabilistic formulation, its integration into an implicit segmentation framework could be realized quite elegantly: A maximum a posteriori estimation of a level set function whose zero level set best separates the organ from the background was formulated under a shape constraint introduced by the GGM-SSM. This led to an energy functional which was optimized in a two-phase formulation alternating a gradient descent with respect to the embedding level set function and the GGM-SSM deformations. The coupling between point-based statistical shape models and level sets is new to our knowledge of the literature and opens new insights on how to take the best of both worlds. From a theoretical point of view, a very powerful feature of the method is that a unique criterion is optimized, thus, the convergence is ensured. Due to the implicit formulation of the approach, new a priori knowledge or constraints can be taken into account as needed for specific applications. This was exemplarily demonstrated by the integration of a boundary term into the energy functional.

As demonstrated further, the segmentation method could be adapted to multiple-object segmentation in a straightforward manner. The shape and location relations of an anatomical structure with regard to their neighbouring structures are interesting information to be used as a-priori knowledge in a segmentation process in order to render the result more robust. For the segmentation algorithm, a separate level set function was defined for each object. Their spatial evolutions during segmentation were then linked and constrained by a single GGM-SSM which models all involved objects in one shape prior. This constitutes another scientific contribution not yet published elsewhere.

Evaluations on kidney data showed that the integration of the shape prior into the level set segmentation offers a solution to the typical implicit segmentation problem of leakage and such brings robustness to the segmentation process. A first evaluation on hip articulation data indicated the well-posedness of the new method to multiple-object segmentation and segmentation of objects featuring non-spherical topology.

6.2   Perspectives

6.2.1   Parameters

The role of the adjustable parameters in both the SSM method and the segmentation method should be subject to further research. Up to now, the parameter values are determined largely heuristically which is not an optimal approach.

SSM Computation: Since the EM-ICP registration is implemented in a multi-scale framework, the three parameters 'initial variance', 'reduction factor' and 'number of iterations' (or final variance respectively) have to be fixed beforehand. The experiments conducted during the research for this thesis suggest that a good choice for the final variance is a value which lies in the order of the squared average point distance of the observations. The choice for the initial variance depends on the shape differences in the training data set. In general, a slower reduction of variance reduces the risk of freezing in a local minimum during optimization. However, in practice a reasonable balance between computational time and that risk has to be found. In theory, these parameters could be modeled in a probabilistic formulation. By doing so, the EM-ICP parameters might become part of the optimization process in the SSM computation and be integrated into the maximum a posteriori estimation presented in chapter 3 as additional observation parameters.

Segmentation: In the segmentation methods, weighting coefficients are employed to control the influence of the different terms in the energy functional as presented in chapter 5. As the energy functional is derived by a MAP explanation, in theory all coefficients should be equal to 1. Expanding on this probabilistic analogy, the traditional coefficients of the variational methods can be seen as powering factors which flatten or peak the density distributions. While the free choice of weights renders the algorithm flexible to different segmentation demands, it also requires a certain user-interaction which should be reduced. This could be done by evaluating the influence of each term and especially the relations between different terms on a set of standard segmentation problems. For example, the experiments conducted in the course of this thesis suggest that a smoothing term becomes obsolete if the SSM term is weighted noticeably.

Furthermore, it would be of interest to investigate an approach were the weights are no longer represented by scalars but by spatial functions. This would allow an adaption of the impact of the respective terms to local image characteristics. Needless to say, the task of defining good weights would become even more complex but it could make sense to try for certain specific applications.

6.2.2   Application

The segmentation method presented in the course of this thesis joins the advantages of explicitly represented shape priors and the advantages of implicit segmentation schemes. The algorithm is therefore very flexible to different kinds of segmentation problems. Especially multiple-object segmentation is of interest as not many approaches exist in that domain. Possible applications are the segmentation of lung and other organs at risk supporting the radiotherapy planning for lung tumors. Typically, the lung movement during inspiration and expiration influences the movement and deformations of the organs lying close by as for example the liver. The new segmentation method offers an easy integration of regional statistics. The grey value distributions of the lung and the grey value distributions of the liver could be sampled and modeled separately. The shape prior on the other hand could comprise the lung and the liver in a single GGM-SSM. By adjusting the influence of the respective terms in the energy functional, the segmentation process can be adapted to the demands of the specific patient's images. For example in images featuring noise or low contrasts, the shape prior term weights could be turned up with respect to the weights of the image information term, so a robust segmentation should be possible. First experiments are currently done in cooperation with the group around J. Ehrhardt from the University Medical Center Hamburg-Eppendorf.

6.2.3   Related Work

For further research in shape modeling it would be worthwhile to study the mathematical relations of the Gaussian mixture model proposed here and the concept of another generative statistical model without one-to-one correspondences as recently proposed by Durrleman et al. [Durrleman 2009]. Similarly to the method presented in this thesis, they interpret the shape observations as randomly generated by the model and formulate the model computation in a maximum a posteriori explanation. However in their approach, the similarity of shapes is measured by a distance on currents that does not assume any type of point correspondences.

   Concerning the segmentation algorithm, an interesting approach was proposed by Raviv et al. [Raviv 2009] which is also developed in a probabilistic framework. An energy functional similar to the one presented in this thesis is optimized for the implicitly represented segmentation contour. However, their approach is designed for group-wise segmentation and chooses a generative method where the unknown segmentation contours are interpreted as randomly generated by the shape prior. As a novelty, the shape prior (described by an atlas) is integrated as an additional unknown parameter which is inferred from the data set through an alternating optimization of the functional. This idea could be extended by replacing the implicitly represented atlas with an explicitly represented SSM which offers a physically interpretable variability model. As the GGM-SSM already is computed in a probabilistic formulation in a generative method, the extension of the segmentation algorithm presented here to a generative segmentation algorithm should be quite direct.

6.2.4 Other

In chapter 4, the problems of the SSM performance measure 'specificity' were illustrated. In general, a fair comparison of different SSM methods is difficult. First, the quality of SSMs is strongly related to the quality of correspondence determination. However, no gold standards for correspondences exist. Secondly, the comparison of SSMs based on different representations is a challenge as most metrics will inherently favor one or the other SSM. In the case shown in this thesis, a surface-based SSM was compared to the point-based GGM-SSM. As a volume overlap metric was deemed to be more meaningful than point distances in the respective experiments, a surface had to be approximated for the GGM-SSM. Naturally, the accuracy of the binary representation then depended on the quality of the approximated surfaces which means that the evaluation results have to be taken with a pinch of salt.

An interesting approach to tackle the problem of finding a correspondence-independent benchmark has recently been proposed by Munsell et al. who introduce a ground truth SSM [Munsell 2008] for 2D evaluation. The proposed benchmark first generates a synthetic training data set by randomly sampling a given SSM that defines a ground-truth shape space. The quality of a new SSM computed on the training data set is evaluated by comparing its shape space against the ground-truth shape space. An extension of the algorithm to 3D SSMs should be straightforward. Furthermore, the approach could be extended to a general framework which also allows an equitable comparison of SSMs based on different representations.

Mathematical Background

A.1 Mathematical Prepositions

Singular Value Decomposition (SVD)

Any real matrix $A \in \mathbb{R}^{m \times n}$ can be decomposed into

$$A = U\Sigma V^T$$

with U being an orthogonal matrix $U \in \mathbb{R}^{m \times m}$, V^T being the transpose of the orthogonal matrix $V \in \mathbb{R}^{n \times n}$ and Σ being a diagonal matrix $\Sigma \in \mathbb{R}^{m \times n}$ with the singular values σ_i on the diagonal in descending order $\sigma_1 \geq \sigma_2 \geq ... \geq \sigma_{min(m,n)}$. This singular values are all non-negative.

However, the number of non-zero values in Σ is less or equal than $min(m,n)$. For the following let us assume $n < m$. By arranging the information given by the SVD in the optimal way we can save a lot of disk space by reducing the matrix dimensions to

$$A = \tilde{U}\tilde{\Sigma}\tilde{V}^T$$

with $\tilde{U} \in \mathbb{R}^{m \times n}$, $\tilde{V} \in \mathbb{R}^{n \times n}$ and $\tilde{\Sigma} \in \mathbb{R}^{n \times n}$.

The singular values and associated pairs of singular vectors u and v of a matrix A satisfy

$$Av_i = \sigma_i u_i$$

and

$$A^T u_i = \sigma_i v_i.$$

In a geometric sense this means that for every rectangular matrix we can find an orthogonal basis V of which each i-th vector v_i is mapped to a non-negative multiplicative of the i-th vector of a orthogonal basis U (if $n > m$ it is $Av_i = 0$ for $i > m$). The singular values σ_i of a matrix A are the square roots of the eigenvalues of $A^T A$.

Eigenvalue Decomposition Using the Jacobi Method

A real symmetric matrix $A \in \mathbb{R}^{n \times n}$ has always real eigenvalues and orthogonal eigenvectors. A can then be written as

$$A = USU^T$$

where $S \in \mathbb{R}^{n \times n}$ is a diagonal matrix which contains the eigenvalues of A on its diagonal, and $U \in \mathbb{R}^{n \times n}$ is composed of the eigenvectors of A and is therefore orthogonal. This formulation of A is called *spectral* or *eigen* decomposition.

In order to calculate the pseudoinverse A^+ for a symmetric matrix, we can use the eigenvalue decomposition instead of the SVD as

$$
\begin{aligned}
AA^+A &= USU^TUS^+U^TUSU^T \\
&= USU^TUS^+SU^T \\
&= USU^TUU^T \\
&= USU^T \\
&= A.
\end{aligned}
$$

The Jacobi method is an iterative algorithm for finding all eigenpairs for a symmetric matrix $A \in \mathbb{R}^{n \times n}$. For small matrices, the Jacobi method gives uniformly accurate results comparable to the QR algorithm. The algorithm determines the sequence of orthogonal matrices $U_1, U_2, ..., U_n$ and the sequence $S_0, S_1, ...$ as follows:

$$
\begin{aligned}
S_0 &= A \\
S_k &= U_k^T S_{k-1} U_k.
\end{aligned}
$$

The sequence $U_1, U_2, ..., U_n$ is constructed in a way that

$$
\lim_{k \to \infty} S_k = S = diag(\lambda_1, \lambda_2, ..., \lambda_n)
$$

with $\lambda_1, \lambda_2, ..., \lambda_n$ being the eigenvalues of A.

The algorithm generates

$$
S_n = U_n^T U_{n-1}^T ... U_1^T A U_1 U_2 ... U_n.
$$

As all U_k are orthogonal, we can write

$$
A = U_1 U_2 ... U_n S_n U_n^T U_{n-1}^T ... U_1^T.
$$

For $n \to \infty$ we obtain $S_n = S$, and hence $U = U_1 U_2 ... U_n$ represents the matrix of eigenvectors of A which gives the eigenvalue decomposition

$$
A = USU^T.
$$

In practice, the algorithm is stopped when the off-diagonal elements of S are close to zero.

The eigenvalue decomposition using the Jacobi method can also be applied to the computation of the pseudo-inverse A^+ of the real symmetric matrix A.

$$
A^+ = US^+U^T.
$$

The computation of S^+ can be done directly by replacing every non-zero entry in S with its reciprocal and then transposing the resulting matrix.

A.2   The ICP as a specific case of the EM-ICP

We want to take a closer look at the computation of the expectation of the correspondence probabilities as defined in equation (3.15). This formulation is numerically unstable, so we reformulate it to

$$
\begin{aligned}
E_{H_{ij}} &= \frac{\exp(-\mu(s_i, T \star m_j))}{\sum_k \exp(-\mu(s_i, T \star m_k))} \\
&= \frac{1}{1 + \sum_{k \neq j} exp(\mu(s_i, T \star m_j) - \mu(s_i, T \star m_k))} .
\end{aligned}
\tag{A.1}
$$

If we assume homogeneous and isotropic Gaussian noise with the variance σ^2, equation (A.1) can be written as

$$
E_{H_{ij}} = \frac{1}{1 + \sum_{k \neq j} exp\left(\frac{(s_i - T \star m_j)^2 - (s_i - T \star m_k)^2}{2\sigma^2}\right)} = \frac{1}{1 + \sum_{k \neq j} r_{ijk}} .
$$

$$
\lim_{\sigma^2 \to 0} r_{ijk} = \begin{cases} 0 & \text{if } (s_i - T \star m_j)^2 < (s_i - T \star m_k)^2 \\ +\infty & \text{if } (s_i - T \star m_j)^2 > (s_i - T \star m_k)^2 \end{cases} .
$$

We see that

$$
\lim_{\sigma^2 \to 0} E_{H_{ij}} = \begin{cases} 1 & \text{if } (s_i - T \star m_j)^2 < (s_i - T \star m_k)^2 \\ 0 & \text{if } (s_i - T \star m_j)^2 > (s_i - T \star m_k)^2 \end{cases}
$$

so the expectation value for the correspondence between two points s_i and m_j is 1 if and only if m_j is the closest point to s_i. For all other points m_k with $k \neq j$ the expectation value of the correspondence becomes 0. This shows that the EM-ICP algorithm behaves like the ICP algorithm for small variances.

A.3   Mathematical Derivations Chapter 3

Derivative of the Second Term for the Global Criterion

By optimizing the global criterion in equation (3.13) alternately with respect to the operands in $\{Q, \Theta\}$, we are able to determine all parameters we are interested in. As some terms recur in the different optimizations, we will introduce the following notations for simplification reasons:
The derivative of the second term of the global criterion is always performed in the same manner. We will demonstrate the application of chain and product rule and then name the resulting terms. The derivative of

$$
\xi_{kij}(T_k, \Omega_k, \bar{M}, v_p, \lambda_p) = \log \sum_{j=1}^{N_m} \exp\left(-\frac{\|s_{ki} - T_k \star m_{kj}\|^2}{2\sigma^2}\right)
$$

with respect to one of the function's parameters (let's say x) is found as follows:

$$\frac{\partial \xi(x)}{\partial x} = \log(u(x))$$

$$= \frac{1}{u(x)} \frac{\partial u(x)}{\partial x},$$

$$u(x) = \sum_{j=1}^{N_m} \exp\left(-\frac{\|s_{ki} - T_k \star m_{kj}\|^2}{2\sigma^2}\right).$$

$$\frac{\partial u(x)}{\partial x} = \sum_{j=1}^{N_m} \exp(f(x)) \frac{\partial f(x)}{\partial x}, \tag{A.2}$$

$$f(x) = -\frac{\|s_{ki} - T_k \star m_{kj}\|^2}{2\sigma^2}.$$

$$\frac{\partial f(x)}{\partial x} = -\frac{\partial}{\partial x} \frac{(s_{ki} - T_k \star m_{kj})^T (s_{ki} - T_k \star m_{kj})}{2\sigma^2}$$

$$= -\frac{(s_{ki} - T_k \star m_{kj})^T}{\sigma^2} \frac{\partial (s_{ki} - T_k \star m_{kj})}{\partial x}$$

So we find the recurring derivative with

$$\frac{\partial \xi}{\partial x} = -\sum_{j=1}^{N_m} \frac{\exp\left(-\frac{\|s_{ki} - T_k \star m_{kj}\|^2}{2\sigma^2}\right)}{\sum_{l=1}^{N_m} \exp\left(-\frac{\|s_{ki} - T_k \star m_{kl}\|^2}{2\sigma^2}\right)} \frac{(s_{ki} - T_k \star m_{kj})^T}{\sigma^2} \frac{\partial (s_{ki} - T_k \star m_{kj})}{\partial x}.$$

By denoting the weight introduced by the correspondence probabilities with

$$\gamma_{ijk} = \frac{\exp\left(-\frac{\|s_{ki} - T_k \star m_{kj}\|^2}{2\sigma^2}\right)}{\sum_{l=1}^{N_m} \exp\left(-\frac{\|s_{ki} - T_k \star m_{kl}\|^2}{2\sigma^2}\right)}$$

the derivative is simply written as

$$\frac{\partial \xi}{\partial x} = -\sum_{j=1}^{N_m} \gamma_{kij} \frac{(s_{ki} - T_k \star m_{kj})^T}{\sigma^2} \frac{\partial (s_{ki} - T_k \star m_{kj})}{\partial x}. \tag{A.3}$$

Optimization with Respect to the Affine Matrix

We have to solve the derivative of the criterion $C_k'(Q_k, \Theta)$ with respect to A_k.

Here, we use the derivative form shown in equation (A.2) and hence differentiate $f(x)$ with respect to A_k:

$$\frac{\partial C_k'(Q_k, \Theta)}{\partial A_k} = -\sum_{i=1}^{N_k} \sum_{j=1}^{N_m} \gamma_{kij} \frac{\partial}{\partial A_k} \frac{\|s_{ki}' - A_k m_{kj}'\|^2}{2\sigma^2}$$

$$\begin{aligned}
\frac{\partial}{\partial A_k} \|si_{ki} - A_k m_{kj}'\|^2 &= \frac{\partial}{\partial A_k} (s_{ki}' - A_k m_{kj}')^T (s_{ki}' - A_k m_{kj}') \\
&= \frac{\partial}{\partial A_k} (s_{ki}'^T s_{ki}' - s_{ki}'^T A_k m_{kj}' - (A_k m_{kj}')^T s_{ki}' - (A_k m_{kj}')^T A_k m_{kj}') \\
&= \frac{\partial}{\partial A_k} (s_{ki}'^T s_{ki}' - s_{ki}'^T A_k m_{kj}' - s_{ki}'^T A_k m_{kj}' + m_{kj}'^T A_k^T A_k m_{kj}').
\end{aligned}$$

Setting the derivative to zero, we find

$$\frac{\partial C_k'(Q_k, \Theta)}{\partial A_k} = 0$$

$$\Leftrightarrow A_k \sum_{i=1}^{N_k} \sum_{j=1}^{N_m} \gamma_{kij} m_{kj}' m_{kj}'^T = \sum_{i=1}^{N_k} \sum_{j=1}^{N_m} \gamma_{kij} s_{ki}' m_{kj}'^T$$

$$\Leftrightarrow A_k \Upsilon_k = \Psi_k, \quad \Upsilon_k, \Psi_k \in \mathbb{R}^{3 \times 3}.$$

Optimization with Respect to the Deformation Coefficients

For the derivative of the second term of the criterion, again the general derivative described in equation (A.3) is employed:

$$\begin{aligned}
\frac{\partial C_k(Q_k, \Theta)}{\partial \omega_{kp}} &= \frac{\omega_{kp}}{\lambda_p^2} + \sum_{i=1}^{N_k} \sum_{j=1}^{N_m} \gamma_{kij} \frac{(s_{ki} - T_k \star m_{kj})^T}{\sigma^2} \frac{\partial(s_{ki} - T_k \star m_{kj})}{\partial \omega_{kp}} \\
&= \frac{\omega_{kp}}{\lambda_p^2} + \sum_{i=1}^{N_k} \sum_{j=1}^{N_m} \gamma_{kij} \frac{(s_{ki} - T_k \star m_{kj})^T}{\sigma^2} \frac{\partial(s_{ki} - t_k - A_k m_{kj})}{\partial \omega_{kp}}.
\end{aligned}$$

As we know $m_{kj} = \bar{m}_j + \sum_{q=1}^{n} \omega_{kq} v_{qj}$ we differentiate

$$\begin{aligned}
\frac{\partial(s_{ki} - t_k - A_k m_{kj})}{\partial \omega_{kp}} &= \frac{\partial}{\partial \omega_{kp}} (s_{ki} - t_k - A_k (\bar{m}_j + \sum_{q=1}^{n} \omega_{kq} v_{qj})) \\
&= -A_k v_{pj}.
\end{aligned}$$

and finally find

$$\frac{\partial C_k(Q_k, \Theta)}{\partial \omega_{kp}} = \frac{\omega_{kp}}{\lambda_p^2} - \frac{1}{\sigma^2} \sum_{i=1}^{N_k} \sum_{j=1}^{N_m} \gamma_{kij} (s_{ki} - T \star m_{kj})^T A_k v_{pj}.$$

Setting $\frac{\partial C_k(Q_k, \Theta)}{\partial \omega_{kp}} = 0$ leaves us with the following three components:

$$0 = \frac{\sigma^2}{\lambda_p^2} \omega_{kp} - \sum_{i=1}^{N_k} \sum_{j=1}^{N_m} \gamma_{kij} (s_{ki} - t_k - A_k \bar{m}_j)^T A_k v_{pj}$$

$$+ \sum_{q=1}^{n} \omega_{kq} \sum_{i=1}^{N_k} \sum_{j=1}^{N_m} \gamma_{kij} v_{qj}^T A_k^T A_k v_{pj}.$$

The solution of this equation with respect to all ω_{kp} is then done by switching to a matrix notation.

A.4   Mathematical Derivations Chapter 5

In this section we present some mathematical rules which were used for the derivatives of the energy terms in section 5.2.3.

A.4.1   Divergence Calculus

We denote $div(V)$ as the divergence of the continuously differentiable vector field V. The divergence in the 3D Euclidian space is defined as the scalar valued function

$$div(V) = \frac{\partial V_x}{\partial x} + \frac{\partial V_y}{\partial y} + \frac{\partial V_z}{\partial z}.$$

The result is invariant under orthogonal transformations.
For several derivative steps in section 5.2.3, we need the following product rule:

$$div(g \cdot V) = g \cdot div(V) + \ <\nabla g, V> \tag{A.4}$$

or in integral form

$$\int_\Omega div(g \cdot V) = \int_\Omega g \cdot div(V) + \int_\Omega \ <\nabla g, V> . \tag{A.5}$$

We denote ∇g as the gradient of the scalar field g. ∇g is a vector field with each vector pointing in the direction of the steepest slope. The steeper the slope, the longer the associated vector.

$$\nabla g = \begin{pmatrix} \frac{\partial g}{\partial x_1} \\ \frac{\partial g}{\partial x_2} \\ \vdots \\ \frac{\partial g}{\partial x_n} \end{pmatrix}.$$

We also know that the integral of the divergence of a vector field equals the projection of that field on the normal vectors n at the edge (the integral of the surface boundary):

$$\int_\Omega div(g \cdot V) = \int_{\partial\Omega} <g \cdot V, n> dn.$$

This means that

$$\int_\Omega g \cdot div(V) + \int_\Omega <\nabla g, V> = \int_{\partial\Omega} <g \cdot V, n> dn. \qquad (A.6)$$

Besides, assuming that there are no objects outside the image, we know that $\int_{\partial\Omega} <g \cdot V, n> dn = 0$ which leaves us in that cases with

$$\int_\Omega g \cdot div(V) = -\int_\Omega <\nabla g, V>.$$

A.4.2 Helpful Derivations

This derivation is used for the differentiation of the shape prior term in section 5.2.3.

$$
\begin{aligned}
|x + \eta y| &= \sqrt{(x + \eta y)^2} \\
&= \sqrt{|x|^2 + 2\eta x^T y + \eta^2 |y|^2} \\
&= |x|\sqrt{1 + 2\eta \frac{x^T y}{|x|^2} + \eta^2 \frac{|y|^2}{|x|^2}} \\
&= |x|(1 + \eta \frac{x^T y}{|x|^2} + O(\eta^2)) \\
&= |x| + \eta \frac{x^T y}{|x|} + O(\eta^2).
\end{aligned}
\qquad (A.7)
$$

The transfer from line 3 to line 4 makes use of a binomial series.

List of Publications

This thesis is a monograph which contains unpublished material. It is however largely based on the following international publications:

Generative Gaussian Mixture Statistical Shape Model (GGM-SSM)

[Hufnagel 2007b]: H. Hufnagel, X. Pennec, J. Ehrhardt, H. Handels, and N. Ayache. *Shape Analysis Using a Point-Based Statistical Shape Model Built on Correspondence Probabilities.* In Proceedings of the Medical Imaging Computing and Computer Assisted Intervention (MICCAI) 2007, volume 4791 of LNCS, pages 959-967, 2007.

[Hufnagel 2007a]: H. Hufnagel, X. Pennec, J. Ehrhardt, H. Handels, and N. Ayache. *Point-Based Statistical Shape Models with Probabilistic Correspondences and Affine EM-ICP.* Bildverarbeitung für die Medizin 2007, Springer Verlag, pages 434-438, 2007.

[Hufnagel 2008b]: H. Hufnagel, X. Pennec, J. Ehrhardt, N. Ayache, and H. Handels. *Generation of a Statistical Shape Model with Probabilistic Point Correspondences and EM-ICP.* International Journal for Computer Assisted Radiology and Surgery (IJCARS) vol. 2, no. 5, pages 265-273, March 2008.

[Hufnagel 2008c]: H. Hufnagel, X. Pennec, J. Ehrhardt, H. Handels, and N. Ayache. *A Global Criterion for the Computation of Statistical Shape Model Parameters Based on Correspondence Probabilities.* Bildverarbeitung für die Medizin 2008, Springer Verlag, pages 277-282, 2008.

Evaluation of the GGM-SSM

[Hufnagel 2008a]: H. Hufnagel, X. Pennec, J. Ehrhardt, N. Ayache, and H. Handels. *Comparison of Statistical Shape Models Built on Correspondence Probabilities and One-to-One Correspondences.* In Proceedings of the SPIE Symposium on Medical Imaging 2008, vol. 6914 of SPIE Conference Series, pages 4T1-4T8, 2008.

[Hufnagel 2009a]: H. Hufnagel, J. Ehrhardt, X. Pennec, N. Ayache, and Heinz Handels. *Computation of a Probabilistic Statistical Shape Model in a Maximum-a-posteriori Framework.* Methods of Information in Medicine, vol. 48, no. 4, pages

314-319, 2009.

Segmentation Using the GGM-SSM

[Hufnagel 2009b]: H. Hufnagel, J. Ehrhardt, X. Pennec, and H. Handels. *Application of a Probabilistic Statistical Shape Model to Automatic Segmentation.* In Proceedings of the World Congress on Medical Physics and Biomedical Engineering 2009, 25/IV, pages 2181-2184, 2009.

[Hufnagel 2009c]: H. Hufnagel, J. Ehrhardt, X. Pennec, A. Schmidt-Richberg, and H. Handels. *Level Set Segmentation Using a Point-Based Statistical Shape Model Relying on Correspondence Probabilities.* In Proceedings of the MICCAI Workshop Probabilistic Models for Medical Image Analysis (PMMIA) 2009, pages 34-44, 2009.

[Hufnagel 2010]: H. Hufnagel, J. Ehrhardt, X. Pennec, A. Schmidt-Richberg, and H. Handels. *Coupled Level Set Segmentation Using a Point-Based Statistical Shape Model Relying on Correspondence Probabilities.* In Proceedings of the SPIE Symposium on Medical Imaging 2010, pages 6914 4T1-4T8, 2010.

Bibliography

[Amenta 1999] N. Amenta and M. Bern. *Surface reconstruction by Voronoi filtering.* Discrete Computational Geometry, vol. 22, pages 481–504, 1999.

[Andreopoulos 2008] A. Andreopoulos and K. Tsotsos. *Efficient and generalizable statistical models of shape and appearance for analysis of cardiac MRI.* Medical Image Analysis, vol. 12, no. 3, pages 335–357, 2008.

[Ballester 2005] M. A. González Ballester, M.G. Linguraru, M. Reyes-Aguirre and N. Ayache. *On the Adequacy of Principal Factor Analysis for the Study of Shape Variability.* In SPIE Medical Imaging, volume 5747, pages 1392–1399, 2005.

[Benameur 2003] S. Benameur, M. Mignotte, S. Parent, H. Labelle, W. Skalli and J. de Guise. *3D/2D registration and segmentation of scoliotic vertebrae using statistical models.* Comput Med Imaging Graph., vol. 27, pages 321–337, Sep-Oct 2003.

[Benayoun 1994] A. Benayoun, N. Ayache and I. Cohen. *Adaptive meshes and nonrigid motion computation.* In International Conference on Pattern Recognition, pages 730–732, 1994.

[Besl 1992] P. J. Besl and N. D. McKay. *A method for registration of 3D shapes.* IEEE Transactions PAMI, vol. 14, pages 239–256, 1992.

[Boisvert 2008] J. Boisvert, F. Cheriet, X. Pennec, H. Labelle and N. Ayache. *Geometric Variability of the Scoliotic Spine using Statistics on Articulated Shape Models.* IEEE Transactions on Medical Imaging,, vol. 27, no. 4, pages 557–568, 2008.

[Bookstein 1986] Fred .L. Bookstein. *Size and shape spaces for landmark data in two dimensions (with discussion).* Statist. Sci., vol. 1, pages 181–242, 1986.

[Bookstein 1991] Fred L. Bookstein. Morphometric tools for landmark data: Geometry and biology. cup, 1991.

[Bookstein 1996] Fred L. Bookstein. *Landmark methods for forms without landmarks: morphometrics in group differences in outline shapes.* Medical Image Analysis, vol. 1, pages 225–243, 1996.

[Bookstein 1997] F.L. Bookstein. *Shape and the information in medical images: a decade of the morphometric synthesis.* Computer Vision and Image Understanding, vol. 66, no. 2, pages 97–118, 1997.

[Brechbühler 1995] C. Brechbühler, G. Gerig and O. Kübler. *Parametrization of closed surfaces for 3-D shape description.* In Computer Vision and Image Understanding, volume 61, pages 154 – 170, 1995.

[Brejl 2000] M. Brejl and M. Sonka. *Object localization and border detection criteria design in edge-based image segmentation: automated learning from examples.* IEEE Transactions on Medical Imaging, vol. 19, no. 10, pages 973–985, 2000.

[Broadhurst 2006] R. Broadhurst, J. Stough, S. M. Pizer and E. L. Chaney. *A Statistical Appearance Model Based on Intensity Quantile Histograms.* In IEEE International Symposium on Biomedical Imaging 2006, pages 422–425, 2006.

[Bronstein 2000] I.N. Bronstein, K.A. Semendjajew, G. Musiol and H. Mühlig. Taschenbuch der Mathematik. Harri Deutsch, 2000.

[Brunelli 2001] R. Brunelli and O. Mich. *Histograms analysis for image retrieval.* Pattern Recognition, vol. 34, no. 8, pages 1625–1637, 2001.

[Caselles 1993] Vicent Caselles, Francine Catté, Tomeu Coll and Françoise Dibos. *A geometric model for active contours in image processing.* Numerical Mathematics, vol. 66, no. 1, pages 1–31, 1993.

[Caselles 1997] V. Caselles, R. Kimmel and G. Sapiro. *Geodesic active contours.* International Journal of Computer Vision, vol. 22, no. 1, pages 61–79, 1997.

[Cates 2006] J. Cates, M. Meyer, P.Th. Fletcher and R. Whitaker. *Entropy-Based Particle Systems for Shape Correspondences.* In Proceedings of the MICCAI'06, volume 1, pages 90–99, 2006.

[Chan 2001] T.F. Chan and L.A. Vese. *Active Contours Without Edges.* IEEE Transactions on Image Processing, vol. 10, no. 2, pages 266–277, 2001.

[Chui 2003] H. Chui, L. Win, R. Schultz, J.S. Duncan and A. Rangarajan. *A unified non-rigid feature registration method for brain mapping.* Medical Image Analysis, vol. 7, pages 113–130, 2003.

[Chui 2004] Haili Chui, Anand Rangarajan, Jie Zhang and Christiana Leonard. *Unsupervised learning of an atlas from unlabeled point-sets.* IEEE Transactions on PAMI'04, vol. 26, pages 160–172, 2004.

[Ciofolo 2005] C. Ciofolo and C. Barillot. *Brain Segmentation with Competetive Level Sets and Fuzzy Control.* In Information Processing in Medical Imaging, volume 3565, pages 333–344, 2005.

[Cootes 1992] T.F. Cootes and C.J. Taylor. *Active shape models – 'smart snakes'.* In British Machine Vision Conference, pages 266–275. Springer, 1992.

[Cootes 1993] T.F. Cootes and C.J. Taylor. *Active shape model search using local grey-level models: A quantitative evaluation.* In British machine Vision Conference, pages 639–648, 1993.

[Cootes 1994] T.F. Cootes and C.J. Taylor. *Using grey-level methods to improve active shape model search.* In International Conference on Pattern Recognition, volume 1, pages 63–67, 1994.

[Cootes 1995] T.F. Cootes, C.J. Taylor, D.H. Cooper and J. Graham. *Active Shape Models - Their Training and Application.* Computer Vision and Image Understanding, vol. 61, pages 38–59, 1995.

[Cootes 1996] T. F. Cootes and C. J. Taylor. *Data driven refinement of active shape model search.* In British Machine Vision Conference, pages 383–392, 1996.

[Cootes 2001a] T. F. Cootes, G. J. Edwards and C. J. Taylor. *Active appearance models.* IEEE Trans Pattern Anal Mach Intell, vol. 23, no. 6, pages 681–685, 2001.

[Cootes 2001b] T. F. Cootes and C. J. Taylor. *On representing edge structure for model matching.* Computer Vision and Pattern Recognition, vol. 1, pages 1114–1119, 2001.

[Cootes 2004] T.F. Cootes and C.J. Taylor. *Statistical Models of Appearance for Computer Vision.* Technical report, University of Manchester, 2004.

[Costa 2007] J. Costa, H. Delingette, S. Novellas and N. Ayache. *Automatic Segmentation of Bladder and Prostate Using Coupled 3D Deformable Models.* In Medical Image Computing and Computer Assisted Intervention (MICCAI'07), pages 252–260, 2007.

[Cremers 2006] Daniel Cremers. *Dynamical statistical shape priors for level set-based tracking.* IEEE Transactions on Pattern Analysis and Machine Intelligence, vol. 28, no. 8, pages 1262–1273, 2006.

[Cremers 2007] D. Cremers, M. Rousson and R. Deriche. *A Review of Statistical Approaches to Level Set Segmentation: Integrating Color, Texture, Motion and Shape.* International Journal of Computer Vision, vol. 72, no. 2, pages 195–215, 2007.

[Davatzikos 1996] C. Davatzikos, M. Vaillant, S.M. Resnick, J.L. Prince, S. Letovsky and R.N. Bryan. *A computerized approach for morphological analysis of the corpus callosum.* Journal of Computer Assisted Tomography, vol. 20, no. 1, pages 88–97, 1996.

[Davies 2002a] R. H. Davies, C. J. Twining, T. F. Cootes, J. C. Waterton and C. J. Taylor. *3D statistical shape models using direct optimisation of description length.* In European Conference on Computer Vision, volume 2352, pages 3–20, 2002.

[Davies 2002b] R.H. Davies. *Learning Shape: Optimal Models of Natural Variability.* PhD thesis, University of Manchester, UK, 2002.

[Davies 2002c] R.H. Davies, C.J. Twining and T.F. Cootes. *A Minimum Description Length Approach to Statistical Shape Modeling.* IEEE Transactions Medical Imaging, vol. 21(5), pages 525–537, May 2002.

[de Bruijne 2003] Marleen de Bruijne, Bram van Ginneken, Max A. Viergever and Wiro J. Niessen. *Adapting Active Shape Models for 3D Segmentation of Tubular Structures in Medical Images.* In Information Processing in Medical Imaging, pages 136–147, 2003.

[de Bruijne 2004] Marleen de Bruijne and Mads Nielsen. *Shape Particle Filtering for Image Segmentation.* In Medical Image Computing and Computer-Assisted Intervention 2004, volume 3216, pages 168–175, 2004.

[de Brujine 2002] M. de Brujine, W.E. Niessen B. van Ginneken, J. Maintz and M. Viergever. *Active shape model based segmentation of abdominal aortic aneurysms in CTA images.* In SPIE Medical Image Processing 2002, volume 4684, pages 463–474, 2002.

[Delingette 2001] H. Delingette and J. Montagnat. *Shape Topology Constraints on Parametric Active Contours.* Computer Vision and Image Understanding, vol. 83, no. 2, pages 140–171, 2001.

[Dempster 1977] A.P. Dempster, N.M. Laird and D.B. Rubin. *Maximum Likelihood from Incomplete Data via the EM Algorithm.* Royal Stat., vol. B 39, pages 1–38, 1977.

[Dervieux 1979] A. Dervieux and F. Thomasset. *A finite element method for the simulation of Raleigh-Taylor instability.* In Springer Lecture Notes in Mathematics, volume 771, pages 145–158, 1979.

[Dey 2004] T.K. Dey and S. Goswami. *Provable Surface Reconstruction from noisy samples.* In Proceedings of the 20th Annual Symposium of Computational Geometry 2004, pages 330–339, 2004.

[Droske 2001] M. Droske, B. Meyer, M. Rumpf and C. Schaller. *An Adaptive Level Set Method for Medical Image Segmentation.* In International Conference on Information Processing in Medical Imaging, volume 2082, pages 416–422, 2001.

[Dryden 1993] I.L. Dryden and K.V. Mardia. *Multivariate Shape Analysis.* The Indien Journal of Statistics, vol. 55, 1993. Dedicated to the memory of P.C. Mahalanobis.

[Duda 1973] R.O. Duda and P.E. Hart. Classification and scene analysis. John Wiley and Sons, 1973.

[Durrleman 2009] Stanley Durrleman, Xavier Pennec, Alain Trouvé and Nicholas Ayache. *Statistical Models on Sets of Curves and Surfaces based on Currents.* Medical Image Analysis, vol. 13, no. 5, pages 793–808, 2009.

[Ericsson 2003] A. Ericsson and K. Aström. *Minimizing the description length using steepest descent.* In British Machine Vision Conference, pages 93–102, 2003.

[Ericsson 2007] Anders Ericsson and Johan Karlsson. *Measures for Benchmarking of Automatic Correspondence Algorithms.* Journal of Mathematical Imaging and Vision, vol. 28, no. 3, pages 225–241, 2007.

[Floreby 1998] L. Floreby, K. Sjogreen, L. Sornmo and M. Ljungberg. *Deformable Fourier Surfaces for Volume Segmentation in SPECT.* In Proceedings of the 14th International Conference on Pattern Recognition, volume 1, pages 358–360, 1998.

[Fogel 1966] L.J. Fogel, A.J. Owens and M.J. Walsh. Artificial intelligence through simulated evolution. John Wiley and Sons, 1966.

[Frangi 2001] Alejandro F. Frangi, Wiro J. Niessen, Daniel Rueckert and Julia A. Schnabel. *Automatic 3D ASM Construction via Atlas-Based Landmarking and Volumetric Elastic Registration.* In Information Processing in Medical Imaging, pages 78–91, 2001.

[Freedman 2005] D. Freedman, R. Radke, T. Zhang, Y. Jeong, D. Lovelock and G. Chen. *Model-based segmentation on medical imagery by matching distributions.* IEEE Transactions on Medical Imaging, vol. 24, no. 3, pages 281–292, 2005.

[Gerig 2001] G. Gerig, M. Styner, M. E. Shenton and J. A. Lieberman. *Shape versus Size: Improved Understanding of the Morphology of Brain Structures.* In Medical Image Computing and Computer-Assisted Intervention - MICCAI 2001, volume 2208, pages 24–32, 2001.

[Golland 2001] P. Golland, W. E. L. Grimson, M. E. Shenton and R. Kikinis. *Deformation Analysis for Shape Based Classification.* In Information Processing in Medical Imaging, volume 2082, pages 517–530, 2001.

[Gonzalez 2002] R.C. Gonzalez and R.E. Woods. Digital image processing. Prentice Hall, Inc., 2002.

[Gotsman 2003] C. Gotsman, X. Gu and A Sheffer. *Fundamentals of spherical parameterization for 3D meshes.* ACM Transactions on Graphics, vol. 22, no. 3, pages 358–363, 2003.

[Granger 2002] S. Granger and X. Pennec. *Multi-scale EM-ICP: A Fast and Robust Approach for Surface Registration*. In Proceedings of the European Conference on Computer Vision 2002, volume 2525 of *LNCS*, pages 418–432, 2002.

[Granger 2003] Sébastien Granger. *Une approche statistique multi-échelle au recalage rigide de surfaces : Application à l'implantologie dentaire*. PhD thesis, Ecole des Mines de Paris, 2003.

[Gu 2003] X. Gu, Y. Wang, T.F. Chan, P.M. Thompson and S.T. Yau. *Genus zero surface conformal mapping and its application to brain surface mapping*. In Information Processing in Medical Imaging, pages 172–184, 2003.

[Handels 2009] Heinz Handels. Medizinische Bildverarbeitung und Mustererkennung - Neue Perspektiven für die bildgestützte Diagnostik und Therapie. Lecture Notes in Informatics. Vieweg und Teubner, 2009.

[Haralick 1985] R.M. Haralick and L.G. Shapiro. *Image Segmentation Techniques*. Graphical Model and Image Processing, vol. 29, pages 100–132, 1985.

[Heimann 2005] T. Heimann, I. Wolf, T. Williams and H.-P. Meinzer. *3D Active Shape Models Using Gradient Descent Optimization of Description Length*. In Proceedings of the Information Processing in Medical Imaging'05, volume 3565, pages 566–577, 2005.

[Heimann 2006] Tobias Heimann, Ivo Wolf and Hans-Peter Meinzer. *Active Shape Models for a Fully Automated 3D Segmentation of the Liver âÄŞ An Evaluation on Clinical Data*. In Medical Image Computing and Computer-Assisted Intervention, volume 4191, pages 41–48, 2006.

[Heimann 2007a] T. Heimann, H.-P. Meinzer and I. Wolf. *A Statistical Deformable Model for the Segmentation of Liver CT Volumes*. In MICCAI 2007 Workshop Proceedings: 3D Segmentation in the Clinic - A Grand Challenge, pages 161–166, 2007.

[Heimann 2007b] T. Heimann, S. Münzing, H.-P. Meinzer and I. Wolf. *A Shape-Guided Deformable Model with Evolutionary Algorithm Initialization for 3D Soft Tissue Segmentation*. In Information Processing in Medical Imaging 2007, volume LNCS 4584, pages 1–12, 2007.

[Heimann 2007c] T. Heimann, I. Wolf and H.-P. Meinzer. *Automatic Generation of 3D Statistical Shape Models with Optimal Landmark Distributions*. Methods Inf Med, vol. 46, no. 3, 2007.

[Heimann 2008] T. Heimann, M. Baumhauer, T. Simpfendörfer, H.-P. Meinzer and I. Wolf. *Prostate segmentation from 3D transrectal ultrasound using statistical shape models and various appearance models.* In SPIE Medical Imaging, volume 6914, pages 69141P1–1P8, 2008.

[Heimann 2009] Tobias Heimann and Hans-Peter Meinzer. *Statistical shape models for 3D medical image segmentation: A review.* Medical Image Analysis, vol. 13, pages 543–563, 2009.

[Hoppe 1992] H. Hoppe, T. DeRose, T. Duchamp, J. McDonald and W. Stuetzle. *Surface reconstruction from unorganized points.* In Proceedings of ACM SIG-GRAPH 1992, volume 26, pages 71–78, 1992.

[Huang 2004] X. Huang, D. Metaxas and T. Chen. *MetaMorphs: Deformable Shapes and Texture Models.* In Computer Vision and Pattern Recognition, volume 1, pages 496–503, 2004.

[Hufnagel 2007a] H. Hufnagel, X. Pennec, J. Ehrhardt, H. Handels and N. Ayache. *Point-Based Statistical Shape Models with Probabilistic Correspondences and Affine EM-ICP.* In Bildverarbeitung für die Medizin 2007, pages 434–438. Springer, 2007.

[Hufnagel 2007b] H. Hufnagel, X. Pennec, J. Ehrhardt, H. Handels and N. Ayache. *Shape Analysis Using a Point-Based Statistical Shape Model Built on Correspondence Probabilities.* In Proceedings of the MICCAI'07, volume 1, pages 959–967, 2007.

[Hufnagel 2008a] H. Hufnagel, X. Pennec, J. Ehrhardt, N. Ayache and H. Handels. *Comparison of Statistical Shape Models Built on Correspondence Probabilities and One-to-One Correspondences.* In Proc. SPIE Symposium on Medical Imaging '08, volume 6914, pages 6914 4T1–4T8, 2008.

[Hufnagel 2008b] H. Hufnagel, X. Pennec, J. Ehrhardt, N. Ayache and H. Handels. *Generation of a Statistical Shape Model with Probabilistic Point Correspondences and EM-ICP.* International Journal for Computer Assisted Radiology and Surgery (IJCARS), vol. 2, no. 5, pages 265–273, 2008.

[Hufnagel 2008c] H. Hufnagel, X. Pennec, J. Ehrhardt, N. Ayache and H. Handels. *A Global Criterion for the Computation of Statistical Shape Model Parameters Based on Correspondence Probabilities.* In Bildverarbeitung für die Medizin 2008, pages 277–282. Springer, 2008.

[Hufnagel 2009a] Heike Hufnagel, J. Ehrhardt, X. Pennec, N. Ayache and H. Handels. *Computing of Probabilistic Statistical Shape Models of Organs Optimizing a Global Criterion.* Methods of Information in Medicine, vol. 48, no. 4, pages 314–319, 2009.

[Hufnagel 2009b] Heike Hufnagel, Jan Ehrhardt, Xavier Pennec and Heinz Handels. *Application of a Probabilistic Statistical Shape Model to Automatic Segmentation.* In World Congress on Medical Physics and Biomedical Engineering, WC 2009, München, pages 2181–2184, 2009.

[Hufnagel 2009c] Heike Hufnagel, Jan Ehrhardt, Xavier Pennec, Alexander Schmidt-Richberg and Heinz Handels. *Level Set Segmentation Using a Point-Based Statistical Shape Model Relying on Correspondence Probabilities.* In Proc. of MICCAI Workshop Probabilistic Models for Medical Image Analysis (PMMIA'09), pages 34–44, 2009.

[Hufnagel 2010] H. Hufnagel, J. Ehrhardt, X. Pennec, A. Schmidt-Richberg and H. Handels. *Coupled Level Set Segmentation Using a Point-Based Statistical Shape Model Relying on Correspondence Probabilities.* In Proc. SPIE Symposium on Medical Imaging 2010, pages 6914 4T1–4T8, 2010.

[Huysmans 2005] T. Huysmans, R. Van Audekercke, J.V. Sloten, H. Bruyninckx and G. Van der Perre. *A three-dimensional active shape model for the detection of anatomical landmarks on the back surface.* Proceedings of the Institution of Mechanical Engineers. Part H, Journal of engineering in medicine, vol. 219, no. 2, pages 129–142, 2005.

[Hyvärinen 2001] A. Hyvärinen, J. Karhunen and E. Oja. *Independent component analysis.* John Wiley and Sons, 2001.

[Kainmüller 2009] D. Kainmüller, H. Lamecker, S. Zachow and H.-C. Hege. *An Articulated Statistical Shape Model for Accurate Hip Joint Segmentation.* In Proc. IEEE Engineering in Medicine and Biology Conference, volume 1, pages 6345–51, 2009.

[Kalman 1996] D. Kalman. *A singular valuable decomposition: The SVD of a matrix.* College Math. Journal, vol. 27, pages 2–23, 1996.

[Kass 1988] M. Kass, A. Witkin and D. Terzopoulos. *Snakes: Active Contour Models.* International Journal of Computer Vision, vol. 1, no. 4, pages 321–331, 1988.

[Kaus 2003] M.R. Kaus, J. von Berg, W. Niessen and V. Pekar. *Automated Segmentation of the Left Ventricle in Cardiac MRI.* In MICCAI 2003, volume LNCS 2878, pages 432–439, 2003.

[Kaus 2004] M.R. Kaus, J. von Berg, J. Weese, W. Niessen and V. Pekar. *Automated segmentation of the left ventricle in cardiac MRI.* Medical Image Analysis, vol. 8, no. 3, pages 245–254, 2004.

[Kelemen 1999] A. Kelemen, G. Szekely and G. Gerig. *Elastic model-based segmentation of 3-D neuroradiological data sets.* IEEE Transactions on Medical Imaging, vol. 18, no. 10, pages 828–839, 1999.

[Kendall 1984] David G. Kendall. *Shape Manifolds, Procrustean Metrics, and Complex Projective Spaces*. Bulletin of the London Mathematical Society, vol. 16, pages 81–121, 1984.

[Kodipaka 2007] S. Kodipaka, B.C. Vemuri, A. Rangarajan, C.M. Leonard, I. Schmallfuss and S. Eisenschenk. *Kernel Fisher discriminant for shape-based classification in epilepsy*. Medical Image Analysis, vol. 11, pages 79–90, 2007.

[Kohlberger 2009] T. Kohlberger, G. Uzunbas, C. Alvino, T. Kadir, D. O. Slosman and G. Funka-Lea. *Organ Segmentation with Level Sets Using Local Shape and Appearance Priors*. In Medical Image Computing and Computer Assisted Intervention, volume 5762, pages 34–42, 2009.

[Kotcheff 1998] A.C.W. Kotcheff and C.J. Taylor. *Automatic construction of eigenshape models by direct optimization*. Medical Image Analysis, vol. 2, no. 4, pages 303–314, 1998.

[Lamecker 2003] H. Lamecker, T. Lange, M. Seebaß, S. Eulenstein, M. Westerhoff and H.-C. Hege. *Automatic Segmentation of the Liver for the Preoperative Planning of Resections*. In Proc. Medicine Meets Virtual Reality (MMVR), pages 171–174, 2003.

[Lamecker 2004] H. Lamecker, M. Seebass, H.-C. Hege and P. Deuflhard. *A 3D Statistical Shape Model of the Pelvic Bone for Segmentation*. In Proc. SPIE Symposium on Medical Imaging, volume 5370, pages 1341–1351, 2004.

[Leventon 2000a] M. Leventon, W. Grimson and O. Faugeras. *Statistical Shape Influence in Geodesic Active Contours*. In Computer Vision and Pattern Recognition, volume 1, pages 316–323, 2000.

[Leventon 2000b] M. Leventon, W. Grimson, O. Faugeras and W. Wells. *Level Set Based Segmentation with Intensity and Curvature Priors*. In IEEE Workshop on Mathematical Methods in Biomedical Image Analysis, MMBIA 2000, pages 4–11, 2000.

[Lin 2006] D.-T. Lin, C.-C. Lei and S.-W. Hung. *Computer-Aided Kidney Segmentation on Abdominal CT Images*. IEEE Transactions on Information technology in Biomedicine, vol. 10, no. 1, pages 59–65, 2006.

[Lorensen 1987] W. E. Lorensen and H. E. Cline. *Marching cubes: A high resolution 3D surface construction algorithm*. In SIGGRAPH 1987: Proceedings of the 14th annual conference on Computer graphics and interactive techniques, pages 163–169, 1987.

[Lorenz 2000] Cristian Lorenz and Nils Krahnstoever. *Generation of Point-Based 3D Statistical Shape Models for Anatomical Objects*. Computer Vision and Image Understanding, vol. 77, pages 175–191, 2000.

[Lorenz 2006] C. Lorenz and J. von Berg. *A comprehensive shape model of the heart.* Medical Image Analysis, vol. 10, no. 4, pages 657–670, 2006.

[Losasso 2006] F. Losasso, R. Fedkiw and S. Osher. *Spatially adaptive techniques for level set methods and incompressible flow.* Computers and Flow, vol. 35, no. 10, pages 995–1010, 2006.

[Lötjönen 2004] J. Lötjönen, S. Kivistö, J. Kokkalainen, D. Smutnek and K. Lauerma. *Statistical shape model of atria, ventricles and epicardium from short- and long-axis MR images.* Medical Image Analysis, vol. 8, no. 3, pages 371–386, 2004.

[Malladi 1995] R. Malladi, J.A. Sethian and B.C. Vemuri. *Shape Modeling with Front Propagation: A Level Set Approach.* IEEE Transactions on Pattern Analysis and Machine Intelligence, vol. 17, no. 2, pages 159–175, 1995.

[Masuda 1996] T. Masuda, K. Sakaue and N. Yokoya. *Registration and Integration of Multiple Range Images for 3-D Model Construction.* In 13th International Conference on Pattern Recognition, volume 1, pages 879–883, 1996.

[McInerney 1996] T. McInerney and D. Terzopoulos. *Deformable models in medical images analysis: a survey.* Medical Image Analysis, vol. 1, pages 91–108, 1996.

[Mitra 2004] N.J. Mitra, A. Nguyen and L. Guibas. *Estimating surface normals in noisy point cloud data.* International Journal of Computational Geometry and Applications (IJCGA), vol. 14, no. 4/5, pages 261–276, 2004.

[Montagnat 2001] J. Montagnat, H. Delingette and N. Ayache. *Review of deformable surfaces: topology, geometry and deformation.* Image and Vision Computing, vol. 19, no. 14, pages 1023–1040, 2001.

[Munsell 2008] B.C. Munsell, P. Dahal and S. Wang. *Evaluating Shape Correspondence for Statistical Shape Analysis: A Benchmark Study.* IEEE Transactions on Pattern Analysis and Machine Intelligence, vol. 30, no. 11, pages 2023–2039, 2008.

[Nain 2007] D. Nain, S. Haker and A. Tannenbaum. *Multi-scale 3-D shape representation and segmentation using sphercial wavelets.* IEEE Transactions on Medical Imaging, vol. 26, no. 4, pages 598–618, 2007.

[Osher 1988] S. Osher and J. Sethian. *Fronts propagation with curvature dependent speed: Algorithms based on Hamilton-Jacobi formulations.* Journal of Computational Physics, vol. 79, pages 12–49, 1988.

[Osher 2003] S. Osher and N. Paragios. Geometric level set methods in imaging, vision, and graphics. Springer, 2003.

[Palm 2001] C. Palm, T.M. Lehmann, J. Bredno, C. Neuschaefer-Rube, S. Klajman and K. Spitzer. *Automated Analysis of Stroboscopic Image Sequences by Vibration Profiles*. In Advances in Quantitative Laryngoscopy, Voice and Speech Research, Procs. 5th International Workshop, pages 1–7, 2001.

[Paragios 2002] N. Paragios and R. Deriche. *Geodesic active regions: A new framework to deal with frame partition problems in Computer Vision*. Journal of Visual Communication and Image Representation, vol. 13, no. 1, pages 249–268, 2002.

[Parzen 1962] Emanuel Parzen. *On Estimation of a Probability Density Function and Mode*. Annals of Mathematical Statistics, vol. 33, no. 3, pages 1065–1076, 1962.

[Paulsen 2003] R.R. Paulsen and K.B Hilger. *Shape modelling using markov random field restoration of point correspondences*. In Information Processing in Medical Imaging, pages 1–12, 2003.

[Pauly 2003] M. Pauly, R. Keiser, L. Kobbelt and M. Gross. *Shape modeling with point-sampled geometry*. In Proceedings of ACM SIGGRAPH 2003, pages 642–650. ACM Press, 2003.

[Pekar 2001] V. Pekar, M.R. Kaus, S. Lobregt C. Lorenz and, R. Truyen and J. Weese. *Shape model based adaptation of 3-D deformable meshes for segmentation of medical images*. In SPIE Medical Imaging, volume 4322, pages 281–289, 2001.

[Pennec 1996] Xavier Pennec. *L'Incertitude dans les Problemes de Reconnaissance et de Recalage - Application en Imagerie Medicale et Biologie Moleculaire*. PhD thesis, Ecole Polytechnique - France, 1996.

[Peter 2006a] A. Peter and A. Rangarajan. *A New Closed-Form Information Metric for Shape Analysis*. In Medical Image Computing and Computer Assisted Intervention (MICCAI) 2006, pages 249–256, 2006.

[Peter 2006b] Adrian Peter and Anand Rangarajan. *Shape Analysis Using the Fisher-Rao Riemannian Metric: Unifying Shape Representation and Deformation*. In IEEE Transactions ISBI'06, pages 1164–1167, 2006.

[Pitiot 2005] A. Pitiot, H. Delingette, P.M. Thompson and N. Ayache. *Expert Knowledge Guided Segmentation System for Brain MRI*. NeuroImage, vol. 23, pages S85–S96, 2005. supplement 1.

[Pizer 1999] S.M. Pizer, P.A. Fritsch D.S.and Yushkevich, V.E. Johnson and E.L. Chaney. *Segmentation, registration, and measurement of shape variation via image object shape*. IEEE Transactions on Medical Imaging, vol. 18, no. 10, pages 851–865, 1999.

[Pizer 2003] S.M. Pizer, P.T. Fletcher, T. Fridmann, D.S. Fritsch, A.G. Gash and et al. *Deformable M-Reps for 3D Medical Image Segmentation*. International Journal of Computer Vision, vol. 55, no. 2-3, pages 85 – 106, 2003.

[Rajamani 2004] K. T. Rajamani, S. C. Joshi and M. A. Styner. *Bone model morphing for enhanced surgical visualization*. In IEEE International Symposium on Biomedical Imaging, volume 2, pages 1255–1258, 2004.

[Rand 1971] William M. Rand. *Objective Criteria for the Evaluation of Clustering Methods*. American Statistical. Association, vol. 66, pages 846–850, 1971.

[Rangarajan 1997a] Anand Rangarajan, Haili Chui and Fred L. Bookstein. *The Soft-assign Procrustes Matching Algorithm*. In Proceedings of the Information Processing in Medical Imaging'97, volume 1230, pages 29–42, 1997.

[Rangarajan 1997b] Anand Rangarajan, Eric Mjolsness, Suguna Pappu, Lila Davachi, Patricia S. Goldman-Rakic and James S. Duncan. *A robust point matching algorithm for autoradiograph alignment*. Medical Image Analysis, vol. 1, no. 4, pages 379–398, 1997.

[Rangarajan 1999] Anand Rangarajan, Haili Chui and James S. Duncan. *Rigid point feature registration using mutual information*. Medical Image Analysis, vol. 3, no. 4, pages 425–440, 1999.

[Raviv 2009] Tammy Riklin Raviv, Koen Van Leemput, William M. Wells III and Polina Golland. *Joint Segmentation of Image Ensembles via Latent Atlase*. In Medical Image Computing and Computer Assisted Intervention (MICCAI'09), pages 272–280, 2009.

[Reyes 2009] M. Reyes, M.A. Gonzalez Ballester, Z. Li, N. Kozic, S. Chin, R.M. Summers and M.G. Linguraru. *Anatomical Variability of Organs via Principal Factor Analysis from the Construction of an Abdominal Probabilistic Atlas*. In IEEE International Symposium on Biomedical Imaging (ISBI), pages 682–685, 2009.

[Rousson 2002] Mikael Rousson and Nikos Paragios. *Shape Priors for Level Set Representations*. In Proceedings of the 7th European Conference on Computer Vision-Part II, 2002.

[Rousson 2004] M. Rousson, N. Paragios and R. Deriche. *Implicit Active Shape Models for 3D Segmentation in MR Imaging*. In Medical Image Computing and Computer-Assisted Intervention - MICCAI 2004, volume 3216, pages 209–216, 2004.

[Sahoo 1988] P.K. Sahoo, S. Soltani, A.K.C. Wong and Y.C. Chen. *A survey of thresholding techniques*. Computer Vision, Graphics, and Image Processing, vol. 41, no. 2, pages 233–260, 1988.

[Säring 2009] Dennis Säring, Jatin Relan, Kai Müllerleile, Michael Groth, Achim Barmeyer, Alexander Stork, Gunnar Lund and Heinz Handels. *Reproducible Extraction of Local and Global Parameters for the Functional Analysis of the Left Ventricle in 4D MR Image Data*. Methods of Information in Medicine, vol. 48, pages 216–224, 2009.

[Schmidt-Richberg 2009] A. Schmidt-Richberg, H. Handels and J. Ehrhardt. *Integrated Segmentation an Non-Linear Registration for Organ Segmentation and Motion Field Estimation in 4D CT Data*. Methods of Information in Medicine, vol. 48, no. 4, pages 344–349, 2009.

[Schroeder 1992] William J. Schroeder, Jonathan A. Zarge and William E. Lorensen. *Decimation of triangle meshes*. Computer Graphics, vol. 26, pages 65–70, 1992.

[Schwefel 1995] H.P. Schwefel. Evolution and optimum seeking. John Wiley and Sons, 1995.

[Seebass 2003] Martin Seebass, Hans Lamecker, Thomas Lange and Peter Wust Johanna Gellermann. *A statistical shape model for segmentation of the pelvic bone*. In ESHO 2003 - 21th Annual Meeting of the European Society of Hyperthermic Oncology, pages 91–93, 2003.

[Sethian 1999] J.A. Sethian. Level set methods and fast marching methods. Cambridge University Press, 1999.

[Shang 2004] Y. Shang and O. Dossel. *Statistical 3D shape-model guided segmentation of cardiac images*. In Computers in Cardiology, pages 553– 556, 2004.

[Soler 2000] L. Soler, H. Delingette, G. Malandain, J. Montagnat and N. Ayache et al. *Fully automatic anatomical, pathological, and functional segmentation from CT scans for hepatic surgery*. In SPIE Medical Image Processing 2000, pages 246–255, 2000.

[Staib 1992] L. H. Staib and J. S. Duncan. *Deformable fourier models for surface finding in 3D images*. In Proc. SPIE Symposium on Medical Imaging 1992, volume 1808, pages 90–104, 1992.

[Staib 1996] L.H. Staib and J.S. Duncan. *Model-based deformable surface finding for medical images*. IEEE Transactions on Medical Imaging, vol. 15, no. 5, pages 720–731, 1996.

[Styner 2001] M. Styner and G. Gerig. *Medial Models Incorporating Object Variability for 3D Shape Analysis*. In Information Processing in Medical Imaging, volume 2082, pages 502–516, 2001.

[Styner 2003a] M. Styner, G. Gerig, J. Lieberman, D. Jones and D. Weinberger. *Statistical shape analysis of neuroanatomical structures based on medial models.* Medical Image Analysis, vol. 7, pages 207–220, 2003.

[Styner 2003b] M. Styner, J.A. Lieberman, D. Pantazis and G. Gerig. *Boundary and Medial Shape Analysis of the Hippocampus in Schizophrenia.* Medical Image Analysis Journal, vol. 8, no. 3, pages 197–203, 2003. Special Issue on MICCAI 2003.

[Styner 2003c] M. Styner, K.T. Rajamani, L.P. Nolte and et al. *Evaluation of 3D Correspondence Methods for Model Building.* In Proceedings for the Information Processing in Medical Imaging'03, volume 2732, pages 63–75, 2003.

[Székely 1996] G Székely, A. Kelemen, C. Brechbühler and G. Gerig. *Segmentation of 2-D and 3-D objects from MRI volume data using constrained elastic deformations of flexible Fourier contour and surface models.* Medical Image Analysis, vol. 1, no. 1, pages 19–34, 1996.

[Terzopoulos 1986] D. Terzopoulos. *Regularization of inverse visual problems involving discontinuities.* IEEE Transactions on Pattern Analysis and Machine Intelligence, vol. 8, no. 4, pages 413–424, 1986.

[Thodberg 2003] H.H. Thodberg. *Minimum description length shape and appearance models.* In Information Processing in Medical Imaging, pages 51–62, 2003.

[Tsaagan 2002] B. Tsaagan, A. Shimizu, H. Kobatake and K. Miyakawa. *An Automated Segmentation Method of Kidney Using Statistical Information.* In MICCAI 2002, volume LNCS 2488, pages 556–563, 2002.

[Tsai 2003] A. Tsai, A. Yezzi, W. Wells, C. Tempany, D. Tucker, A. Fan, W.E. Grimson and A. Willsky. *A Shape-Based Approach to the Segmentation of Medical Imagery using Level Sets.* IEEE Transactions on Medical Imaging, vol. 22, no. 2, pages 137–154, 2003.

[Tsai 2004] A. Tsai, W. Wells, C. Tempany, E. Grimson and A. Willsky. *Mutual information in coupled multi-shaped model for medical image segmentation.* Medical Image Analysis, vol. 8, pages 429–445, 2004.

[Tsai 2005] Andy Tsai, William M. Wells, Simon K. Warfield and Alan S. Willsky. *An EM algorithm for shape classification based on level sets.* Medical Image Analysis, vol. 9, pages 491–502, 2005.

[Vincent 1991] L. Vincent and P. Soille. *Watersheds in digital spaces: an efficient algorithm based on immersion simulations.* IEEE Transactions on Pattern Analysis and Machine Intelligence, vol. 33, no. 13, pages 583–598, 1991.

[Vos 2004] F.M. Vos, P.W de Bruin, G.I. Streeksa, M. Maas, L.J. van Vliet and A.M. Vossepoel. *A statistical shape model without using landmarks*. In Proceedings of the ICPR'04., volume 3, pages 714–717, 2004.

[Wang 2000] Y. Wang and L.H. Staib. *Boundary Finding with Prior Shape and Smoothness Models*. IEEE Transactions on Pattern Analysis and Machine Intelligence, vol. 22, no. 7, pages 738–743, 2000.

[Weese 2001] J. Weese, M. Kaus, Lorenz C, S. Lobregt, R. Truyen and V. Pekar. *Shape Constrained Deformable Models for 3D Medical Image Segmentation*. In Information Processing in Medical Imaging 2001, pages 380–387, 2001.

[Westin 1998] C.F. Westin, S. Warfield, L. Bhalero, L. Mui, J. Richolt and R. Kikinis. *Tensor Controlled Local Structure Enhancement of CT Images for Bone Segmentation*. In Medical Image Computing and Computer Assisted Intervention, pages 1205–1212, 1998.

[Zeng 1999] X. Zeng, L.H. Staib, R.T. Schultz and J.S. Duncan. *Segmentation and measurement of the cortex from 3-D MR images using coupled-surfaces propagation*. IEEE Trans Med Imaging, vol. 18, no. 10, pages 927–937, 1999.

[Zhang 1994] Z. Zhang. *Iterative point matching for registration of free-form curves and surfaces*. International Journal of Computer Vision, vol. 13, no. 2, pages 119 – 152, 1994.

[Zhao 1996] H.-K. Zhao, T. Chan, B. Merriman and S. Osher. *A variational level set approach to multiphase motion*. Journal of Computational Physics, vol. 127, no. 1, pages 179–195, 1996.

[Zhao 2005a] Z. Zhao, S. R. Aylward and E. K. Teoh. *A Novel 3D Partitioned Active Shape Model for Segmentation of Brain MR Images*. In Medical Image Computing and Computer Assisted Intervention (MICCAI) 2005, pages 221–228, 2005.

[Zhao 2005b] Z. Zhao and E. K. Theo. *A novel framework for automated 3D PDM construction using deformable models*. In Proc. of SPIE, Medical Imaging 2005, volume 5747, pages 303–314, 2005.

Glossary